Histoires de la Terre

FAUX TITRE

322

Etudes de langue et littérature françaises
publiées sous la direction de

Keith Busby, M.J. Freeman,
Sjef Houppermans et Paul Pelckmans

Histoires de la Terre

Earth Sciences and French Culture 1740-1940

Edited by

Louise Lyle and David McCallam

Rodopi

AMSTERDAM - NEW YORK, NY 2008

Illustration cover / Illustration couverture:
Detail from a 1944 War Office geological map of the Paris region.

Cover design / Maquette couverture: Pier Post.

The paper on which this book is printed meets the requirements of 'ISO 9706: 1994, Information and documentation - Paper for documents - Requirements for permanence'.

Le papier sur lequel le présent ouvrage est imprimé remplit les prescriptions de 'ISO 9706: 1994, Information et documentation - Papier pour documents - Prescriptions pour la permanence'.

ISBN: 978-90-420-2477-9

Printed in The Netherlands

Table of Contents

List of Contributors

Benoît de Baere is a Research Fellow in the French Department of the University of Ghent in Belgium.

Ceri Crossley is Professor of Nineteenth-Century Studies in the Department of French Studies, University of Birmingham, UK.

Ben Fisher is a Senior Lecturer in French at the University of Bangor, UK.

Rebecca Ford is a Lecturer in the Department of French and Francophone Studies at the University of Nottingham, UK.

Martin Hurcombe is a Senior Lecturer in French at the University of Bristol, UK.

Greg Kerr is currently completing his PhD, supported by the Irish Research Council for the Humanities and Social Sciences, on Saint-Simonian Aesthetics, at Trinity College Dublin.

Claire Le Guillou is an Affiliated Researcher at the Centre de Recherches Révolutionnaires et Romantiques, Université de Clermont-Ferrand, France.

Louise Lyle is a Lecturer in French at the University of London Institute in Paris (ULIP), France.

David McCallam is a Senior Lecturer in the Department of French, University of Sheffield, UK.

Anca Mitroi is an Assistant Professor in French at Brigham Young University, USA.

Grégory Quenet is a *maître de conférences* in Modern History at the Université de Versailles St-Quentin-en-Yvelines, France.

Ian D. Rotherham is a Reader in Environmental Studies at Sheffield Hallam University, UK.

Scott Sprenger is an Associate Professor of French Studies at Brigham Young University, USA.

Tim Unwin is a Professor of French at the University of Bristol, UK.

Kiera Vaclavik is a Lecturer in French at Queen Mary, University of London, UK.

David H. Walker is a Professor in the Department of French at the University of Sheffield, UK.

Acknowledgements

This volume contains the expanded proceedings of a conference entitled *Histoires de la Terre* which took place at the University of Sheffield, 30 March to 1 April 2007. The editors would like to express their sincere thanks to all who offered their support and assistance to the venture. The conference was generously supported by the Society for French Studies and the French Embassy in the United Kingdom, as well as by various bodies within the University of Sheffield, notably the Department of French, the Humanities Division and the Centre for Nineteenth-Century Studies. We are also indebted to Dr Michael Meredith of the University's Humanities Research Institute for his invaluable help in preparing the manuscript for publication.

Introduction

Louise Lyle and David McCallam

This book is based on the proceedings of a highly original international conference entitled *Histoires de la Terre* which took place at the University of Sheffield from 30 March to 1 April 2007. Its chapters explore how Enlightenment and post-Enlightenment developments in the earth sciences and related fields (palaeontology, mining, agronomy, archaeology, seismology, oceanography, evolutionary theory, etc.) impacted on contemporary French culture. They reveal that geological ideas were a much more pervasive and influential cultural force than has hitherto been supposed. From the mid-eighteenth century, with the publication of Buffon's seminal *Théorie de la Terre* (1749), until the early twentieth century, concepts and figures drawn from the earth sciences inspired some of the most important French philosophers, novelists, political theorists, historians and popularizers of science of the time.

The chapters follow a broadly chronological approach and are divided into four sections. These sections deal successively with:

i.) the tentative beginnings of earth science as a discipline in mid-to-late-eighteenth-century France, examining the fascinating divergences between theory and practice, ideas and experience, as well as exploring how such new conceptualizations were presented to an increasingly knowledgeable public;

ii.) how early nineteenth-century thinkers and writers capitalized on the rhetorical and figurative scope of contemporary earth sciences in order to elaborate their often utopian political projects or imaginative fictions, especially in the popular genre of the novel;

iii.) how the latter part of the nineteenth century greatly intensified and diversified the exploitation of the discourses of the earth sciences, with geological figures informing very different aspects of French culture, from children's literature to avant-garde aesthetics, culminating in the popular science-fiction of Jules Verne;

iv.) the early twentieth century marked a further shift in the relation between earth sciences and broader French culture, a shift that encouraged writers, specifically the figure of the "intellectuel", on the one hand to engage critically with the latest technology employed to exploit the earth's resources and on the other to problematize any ideological attachment to the "soil" of *la patrie*.

This is clearly not a history of the development of earth sciences in France or in western Europe since the mid-eighteenth century. Excellent studies on this have already been written by Martin J. S. Rudwick, David Oldroyd, Gabriel Gohau and Barbara Kennedy among others.[1] It is, rather, a cultural history of the uses to which earth sciences, their concepts, figures and fields of knowledge have been put by contemporary French writers and thinkers. It does, however, employ historical parameters: roughly speaking, 1740 to 1940. The reason for this chronological range is that it marks out a paradigm in French thinking about the Earth. It starts with the conceptual move away from the speculative mechanist models of Descartes and his followers in understanding geological phenomena towards the so-called "Neptunist" theories of writers such as Buffon, and the less well-

[1] See in particular Martin J. S. Rudwick's magisterial *Bursting the Limits of Time: The Reconstruction of Geohistory in the Age of Revolution* (Chicago; London: The University of Chicago Press, 2005), as well as his *The New Science of Geology: Studies in The Earth Sciences in the Age of Revolution* (Aldershot: Ashgate, 2004); see also David R. Oldroyd, *Thinking about the Earth: A History of Ideas in Geology* (London: Athlone, 1996), Gabriel Gohau, *Histoire de la géologie* (Paris: Éditions de la Découverte, 1987), and Barbara Kennedy, *Inventing the Earth: Ideas on Landscape Development since 1740* (Malden, MA; Oxford: Blackwell, 2006).

known, Benoît de Maillet, in the late 1740s.[2] These theories, and the "Vulcanist" successors that were to challenge them from the 1760s onwards, replace abstract mechanical models of geological formation with an emphasis on the natural forces shaping the Earth, be they watery (Neptunist) or fiery (Vulcanist), focussing specifically on what Buffon would consider to be "actual causes", that is, observable geological processes active in the present world.[3] This conceptual break is reinforced with the explosion of both theorizing and fieldwork that accompanied the Lisbon earthquake of 1755, a scientific obsession with all things telluric which, if anything, intensified in the following decades. The upper limit of our chronological range – 1940 – approximately marks a further shift in geological thinking, this time away from the paradigm established on the basis of Enlightenment and post-Enlightenment work in the earth sciences to the widespread discussion and ultimate acceptance of such key theories of twentieth- and twenty-first-century geology as continental drift, plate tectonics and the radioactivity of the Earth's core.[4]

Yet it is also worth noting that while we have periodized or historicized our study of the contribution of earth sciences to French culture from the mid-eighteenth century to the early twentieth century, many of the chapters make it abundantly clear that these earth sciences constitute in themselves a fundamental *historicizing* force for their contemporaries. That is, they provide an optic through which to revise traditional chronologies, specifically those of established religion. What is more, the vast temporal panoramas laid out by geology do much more than reveal a past more distant than was ever imagined before; as certain chapters here show, they also serve to redefine perceptions of the historical present (the Lisbon earthquake is a key moment in this process) and to shape anticipations of the future. The implicit or explicit attacks on religion are one of the most salient features of the historicizing perspective afforded by the early earth sciences of the late eighteenth and early nineteenth centuries. Thus the

---

[2] De Maillet's *Telliamed* (Amsterdam: 1748), whose title is his name spelt backwards, contains wildly speculative visions of a universal, primordial sea and a chronology of the Earth stretching back millions of years, both notions that Buffon would develop more circumspectly and empirically.

[3] See Rudwick, *Bursting the Limits of Time*, pp. 140, 178.

[4] See for instance Oldroyd, *Thinking about the Earth*, pp. 248-82.

revolutionary land management of pioneering agronomists such as Arthur Young or the La Rochefoucauld brothers implies a human mastery of the natural environment which has surmounted forever the cyclical fatality of feast-and-famine found in Catholic teachings and familiar to the peasant mindset. More explicitly, the anti-clericalism of the *Encyclopédie*'s articles and plates on geological phenomena only refute the centrality of humankind in the Biblical chronology the better to install a free-thinking humanity at the heart of a new vision of man's relationship to the Earth, a vision which is variously deist, theist or atheist-materialist in tenor depending on the convictions of the author or draughtsman. What the early nineteenth-century utopians, such as the Saint-Simonians and Leroux, take from this tradition is a sense of organicity and oneness with the Earth which they subsequently reimbue with their own deeply anti-Christian mysticism or religiosity.

If the figures drawn from earth sciences by French social utopianism already represent a large rhetorical remove from Enlightenment empiricism, this abstraction and extension of geological ideas is taken further by novelists such as Balzac and Sand who specifically structure psychological and historical narratives on the model of archaeological excavation. Whether in theory or practice, an engagement with archaeology literally informs narratives of national historical rupture that traumatize the fictional subject (Balzac) or tales of local historical continuity that reassure their protagonists (Sand). Later in the century, it is rather the Enlightenment use of the earth sciences as a rationalist tool for the secular education of the young that inspires a generation of writers sharing a fascination for subterranean spaces and figures in their work. In late nineteenth-century France, the young mind – now conceived as a sort of "prehistory" of the adult psyche – is not an already formed structure to be unearthed and analysed dispassionately; instead, it is a secretive world in motion, an active, moulding, shaping, dynamic underground of thoughts, emotions and actions. Fittingly in the age of Darwin, it is a world of fantastical and recapitulative evolutions. Yet, as the stories of Verne, Malot, Ballantyne or MacDonald show us, a profound ambivalence haunts these later geological discourses: subterranean culture can just as easily deform and disfigure as it can form and mould aright.

It is the sheer inscrutability of underground (or underwater) processes that makes them ideal spaces for both physical and metaphysical exploration, from eighteenth-century electrical theories of earthquake formation to Jarry's pataphysics. The earth sciences are the site of both exhaustive empirical fieldwork and the wildest abstract speculation, affording French writers and thinkers a key transitional conceptual space in which to work, accommodating what we might call *parasciences* that allow their practitioners to rethink the human, indeed to think beyond the human. For the geo-sciences are also the cradle of a French ecological consciousness in this period. If, as these chapters suggest, the Earth is a text, it is one inscribed with a persistent anti-anthropocentrism, from Buffon's vision of a irreversibly cooling globe to the irresistible rise of the "ferromagnétaux" in Rosny's science-fiction. The human is but one species among many hosted by the Earth, and is unique only in pursuing an exploitation of the planet's natural resources that might hasten its own extinction. These texts, then, are biocentric, giving equal weight to all life-forms, the more radical of them postulating the ultimate revenge of the mineral realm over the animal and vegetal kingdoms.

Nonetheless, in the early decades of the twentieth century, when the human race seemed increasingly bent on self-destruction, the earth sciences offer French writers such as Gide and Malraux telling metaphors with which to stake out their ideological positions. Hence Gide opposes empirical experiences of agricultural reform and intellectual nomadism to the blood-and-soil rhetoric of extreme right-wing ideologues such as Barrès. As for Malraux, the Earth becomes an apposite figure of the gruff authenticity and noble endurance of both peasantry and the Resistance, articulating in part his flight from Communism to Gaullism in this period. And, even though the Second World War marks the beginning of a paradigm shift in geological thinking, earth science metaphors of an earlier era persist in French culture in general, and in French politics in particular, as is evidenced by terms such as "paysage politique", "raz-de-marée électoral", "la déferlante médiatique", "la faille démocratique", "les couches populaires", "un gisement d'emplois", etc., or most spectacularly the recurrent use of "séisme" to describe socio-political shocks such as Jean-Marie Le Pen's success in the 2002 presidential elections.[5] Hence, the diverse studies in this book present original insights into the formation and

[5] For a couple of examples among many, see the editions of *Le Monde* for 23 and 27 April 2002.

development of a much neglected, yet highly influential, seam of French cultural history.

SECTION 1

THE ENLIGHTENMENT

Natural Catastrophe in Buffon's *Histoire naturelle*: Earth, Science, Aesthetics, Anthropology

Benoît de Baere

Abstract: *Buffon's manifest interest in all forms of natural disaster embraces three main discursive approaches: that of the pioneering earth scientist, that of the* philosophe *focussing on the human experience of natural catastrophe, and that of the aesthete getting to grips with the sublime character of nature's most violent phenomena. Yet these three discourses are far from being mutually compatible. Taking earthquakes, volcanoes and tidal waves as primary examples, this chapter examines the ways in which the* Histoire naturelle *attempts to reconcile its discordant "voices" into a unique, totalizing vision of the natural world.*

Throughout the weighty in-quarto volumes of Buffon's *Histoire naturelle* the reader encounters again and again the fascination that the French naturalist felt for natural catastrophes – for those earthquakes, volcanic eruptions and floods which have shaped the face of the Earth and determined the fate of life on our planet. These events are, however, approached from a number of different angles, for Buffon is not content to provide explanations, he also expresses his opinion on the attitudes which mankind should adopt when confronted with danger on this scale, and in doing so his writing betrays the particular aesthetic fascination that the violence and brutality of unbridled natural power aroused in him.

In this article, it is my intention to study three divergent but complementary approaches to natural catastrophe, those of earth science, aesthetics and anthropology. I will, of course, examine their specific characteristics, but I will also identify the "links" that bind them to each other. In conclusion, I will attempt to show that these three approaches are not entirely compatible, at least as far as the *Histoire naturelle* is concerned. Indeed, despite the fact that Buffon conceived of his work as an integrated whole, as a complete philosophical entity combining a logic, physics, metaphysics and ethics of its own,[1] his treatment of natural catastrophe allows us to assess the degree to which Buffon's thought at times follows paths so divergent that the overall coherence of the *Histoire naturelle* appears compromised.

Natural Catastrophe as Conceived in Buffon's Scientific Thought: from the Théorie de la Terre *to the* Époques de la nature

We ought to begin by stating that Buffon is very well informed on the subject of natural disasters. In fact, the reader quickly gathers the impression that he had at his disposal a veritable inventory in which innumerable natural events were catalogued in terms of their date, place, consequences and casualties. The *Histoire naturelle* contains whole pages of lists which seem to have been drawn up on the basis of just such a catalogue. For the most part, these earthquakes and volcanic eruptions are presented in a neutral, detached tone; references to death tolls are not systematized. This is mainly because Buffon is interested first and foremost in the consequences of these phenomena for a "theory of the Earth", that is, for understanding the formation of the geophysical relief of the world around us. Thus he suggests repeatedly that the number and scale of these cataclysms offer a possible explanation for the uneven nature of the globe's surface. Does this surface not present us with "l'image [...] d'un amas de débris", "d'un monde en ruine"?

> Commençons [...] par nous représenter ce que l'expérience de tous les tems & ce que nos propres observations nous apprennent au sujet de la terre. Ce globe immense nous offre à la surface, des hauteurs, des profondeurs, des plaines, des mers, des marais, des fleuves, des cavernes, des gouffres, des volcans, & à la première inspection nous ne découvrons en tout cela aucune régularité, aucun ordre. Si nous pénétrons dans son intérieur, nous y trouvons des métaux, des minéraux, des pierres, des bitumes, des sables, des terres, des eaux & des

[1] See also Thierry Hoquet, *Buffon: histoire naturelle et philosophie* (Paris: Champion, 2004), pp. 38, 40.

> matières de toute espèce, placées comme au hasard & sans aucune règle apparente; en examinant avec plus d'attention, nous voyons des montagnes affaissées, des rochers fendus & brisez, des contrées englouties, des isles nouvelles, des terreins submergez, des cavernes comblées; nous trouvons des matières pesantes souvent posées sur des matières légères, des corps durs environnez de substances molles, des choses sèches, humides, chaudes, froides, solides, friables, toutes mêlées & dans une espèce de confusion qui ne nous présente d'autre image que celle d'un amas de débris & d'un monde en ruine.[2]

In this extract Buffon takes up – while also embroidering on it – a commonplace of the "earth science" tradition, one that was common currency since at least Thomas Burnet's *Telluris theoria sacra* of 1681. In this very successful work, Burnet opined that if the observation of sea and ocean floors, underground caverns and mountains has anything at all to teach us, it is that our planet in no way re sembles an "assemblage bien ajusté et charmant" ("*ordinata et venusta rerum compages*").[3] On the contrary, he claimed "Our Cities are built upon Ruins, and our fields and Countries stand upon broken Arches and Vaults".[4]

In France this idea had been picked up by Fontenelle, among others. In a frequently cited passage of his *Histoire de l'Académie* from 1718, he makes reference to "les ruines de la croûte extérieure" of the globe. Moreover, he unequivocally associates these ruins with the "grandes revolutions" visited upon the surface of the Earth. He writes:

> Des vestiges très-anciens & en très-grand nombre, d'inondations qui ont dû être très-étendues, & la manière dont on est obligé de concevoir que les montagnes se sont formées, prouvent assez qu'il est arrivé autrefois à la surface de la terre de grandes révolutions. Autant qu'on en a pû creuser, on n'a presque vû que des ruines, des débris, de vastes décombres entassez pêle-mêle, & qui par une longue suite de siècles se sont incorporez ensemble & unis en une seule masse le plus qu'il a été possible; s'il y a dans le globe de la terre quelque espèce d'organisation régulière, elle est plus profonde & par conséquent nous sera

[2] Georges Louis Leclerc, comte de Buffon, "Second discours: histoire & théorie de la terre", in *Histoire naturelle* (Paris: Imprimerie royale, 1749), I, pp. 68-69.

[3] Thomas Burnet, *T. Burnetii telluris theoria sacra, originem et mutationes generales orbis nostri, quas aut iam subiit, aut olim subiturus est, complectens* (Amsterdam: Wolters, 1694), p. 33.

[4] Thomas Burnet, *The Sacred Theory of the Earth: Containing an Account of the Original of Earth, and of all the General Changes which it hath already undergone, or is to undergo, till the Consummation of all Things*, 6th ed. (London: Hooke, 1726) p. 164.

> toûjours inconnue, & toutes nos recherches se termineront à fouiller dans les ruines de la croûte extérieure, elles donneront encore assez d'occupation aux Philosophes.[5]

At first sight, Buffon seems to agree with this argument: "il n'est pas possible de douter", he writes, that there has occurred "une infinité de révolutions, de bouleversemens, de changemens particuliers & d'altérations sur la surface de la terre", citing as examples the "mouvemens naturel des eaux de la mer", "l'action des pluies, des gelées, des eaux courantes, des vents, des feux soûterrains, des tremblemens de terre &c.".[6] All these "particular"[7] causes have not insubstantially "contribué à changer la face du globe"[8] by occasioning "des bouleversemens, des inondations, des affaissemens [etc.]". In short, despite being "ce que nous connoissons de plus solide", the Earth's surface is subject to a series of "vicissitudes perpétuelles".[9] Yet what then is the exact place that these catastrophic events occupy in Buffon's theory of the Earth? What precisely is the role he assigns to them?

The response to be found in the *Histoire naturelle* is much more complex than one might expect. This complexity is manifest right from start of the *Théorie de la Terre* (contained in the first volume of the *Histoire naturelle*), since Buffon urges the reader not to give his or her opinion too "precipitately" on the subject of the "irrégularité que nous voyons à la surface de la terre, & sur le désordre apparent qui se trouve dans son intérieur". Is it not possible that by paying more attention we might discover "un ordre que nous ne soupçonnions pas, & des rapports généraux que nous n'apercevions pas au premier coup d'œil"?[10] We must therefore seek out this "organisation régulière", "plus profonde", whose existence was suspected

---

[5] Bernard Le Bovier de Fontenelle, *Histoire de l'académie royale des sciences de Paris. Année 1718* (Paris: Imprimerie royale, 1720), p. 3. Buffon knew of this passage (see *Histoire naturelle*, I, pp. 304-5) as well as Burnet's work (see the third article of his "Preuves de la théorie de la terre", entitled "Du système de M. Burnet", in *Histoire naturelle*, I, p. 180).

[6] Buffon, "'Conclusion' des Preuves de la théorie de la terre", *Histoire naturelle*, I, p. 611.

[7] Buffon, "Preuves de la théorie de la terre, article XIX: Des changemens de terres en mers, & de mers en terres", *Histoire naturelle*, I, p. 609.

[8] Buffon, "Second discours", *Histoire naturelle*, I, pp. 109-10.

[9] Buffon, "Des changemens de terres en mers", *Histoire naturelle*, I, p. 609.

[10] Buffon, "Second discours", *Histoire naturelle*, I, pp. 69-70.

by Fontenelle but which he believed to be beyond our understanding. This is precisely what Buffon does when he sets out to "représenter ce que l'expérience de tous les temps & ce que nos propres observations nous apprennent au sujet de la Terre".[11] To realize this "représentation", however, Buffon establishes a hierarchy within the observational data at his disposal. All are not equally important, he says, and without hesitation he distinguishes the facts he deems to be "normaux" from events he considers to be "anormaux", that is, the "accidents".

At the end of his inspection of what he calls the "principaux faits" of Earth history, the savant from Montbard states that all "important" phenomena can be explained *exclusively* by the action of regular and continuous causes. What should we make, for example, of that "espèce d'organisation de la Terre que nous découvrons partout, [de] cette situation horizontale & parallèle des couches"? Is it not obvious that it can only come from "une cause constante & et d'un mouvement réglé & toujours dirigé de la même façon"?[12] The "arrangement", as he calls it, must therefore have been produced "par les eaux ou plûtôt par les sédimens qu'elles ont déposez dans la succession des temps; toute autre révolution, tout autre mouvement, toute autre cause auroit produit un arrangement très-différent".[13] Conversely, he has not encountered a single state of affairs that is the necessary result of catastrophic or discontinuous causes. Hence, he writes, it is not *necessary* to explain the "inégalités du globe" in terms of cataclysms; we might just as well suppose them to be the erosive work of heavy rains or the movement of the seas, to cite but two alternative causes.[14] It is only a question of extending sufficiently the duration of such actions. Why, in effect, should we posit extraordinary causes for these features, if everyday phenomena suffice to explain them?

This is the reason why Buffon, throughout the *Discours* that he devotes to Earth theory and Earth history, refrains from invoking "ces causes éloignées qu'on prévoit moins qu'on ne les devine". For

[11] Buffon, *Histoire naturelle*, I, p. 68.
[12] Buffon, *Histoire naturelle*, I, p. 81.
[13] Buffon, "Preuves de la théorie de la terre, article VIII: Sur les coquilles & les autres productions de la mer, qu'on trouve dans l'intérieur de la Terre", *Histoire naturelle*, I, p. 304.
[14] Buffon, "Conclusion", *Histoire naturelle*, I, p. 610.

him, the historian must at all costs avoid mentioning phenomena whose effects are “rare, violent & subit”. Only those “effets qui arrivent tous les jours”, “les mouvemens qui se succèdent & se renouvellent sans interruption”, and those “opérations constantes & toujours réitérées” are worthy of his attention. As he writes: “ce sont là nos causes & nos raisons”.[15] The “system”, so to speak, that Buffon deduced from all of this is summed up in the following passage:

> [C]e sont les courans de la mer qui ont creusé les vallons & élevé les collines en leur donnant des directions correspondantes; ce sont ces mêmes eaux de la mer, qui en transportant les terres, les ont disposées les unes sur les autres par lits horizontaux, & ce sont les eaux du ciel qui peu à peu détruisent l’ouvrage de la mer, qui rabaissent continuellement la hauteur des montagnes, qui comblent les vallées, les bouches des fleuves & les golfes, & qui ramenant tout au niveau, rendront un jour cette terre à la mer, qui s’en emparera successivement, en laissant à découvert de nouveaux continens entre-coupés de vallons & de montagnes, & tout semblables à ceux que nous habitons aujourd’hui.[16]

It is worth noting, however, that Buffon admits that *in theory* it is *possible* that certain natural catastrophes (earthquakes, storms, volcanic eruptions) may play a part in forming the Earth’s surface: “Il n’y auroit [...] *pas d’impossibilité absolue* à supposer que les montagnes ont été élevées par des tremblemens de terre’,[17] and he goes so far as to advance a calculation to prove it.[18] Thus, if Buffon objects to cataclysmic causes intervening in his Earth history, it is solely for the sake of economy: his observations do not require that he take these phenomena into consideration, since ultimately “il est evident pour tous les gens qui se donneront la peine d’observer que l’arrangement de toutes les matières qui composent le globe, est l’ouvrage des eaux”.[19] Here is one of his most explicit statements to this effect:

> Ces énormes ravages produits par les tremblemens de terre ont fait croire à quelques Naturalistes que les montagnes & les inégalités de la surface du globe n’étoient que le résultat des effets de l’action des feux soûterrains, & que toutes les irrégularités que nous remarquons sur la terre, devoient être attribuées à ces secousses violentes & aux bouleversemens qu’elles ont produits; c’est, par

---

[15] Buffon, “Second discours”, *Histoire naturelle*, I, p. 98.
[16] Buffon, *Histoire naturelle*, I, p. 124.
[17] Buffon, “Preuves de la théorie de la terre, article XVI: Des volcans & des tremblemens de terre”, *Histoire naturelle*, I, pp. 523-24. My italics.
[18] Buffon, *Histoire naturelle*, I, p. 524.
[19] Buffon, “Sur les coquilles”, *Histoire naturelle*, I, p. 303.

> exemple, le sentiment de Ray's [*sic*], il croit que toutes les montagnes ont été formées par des tremblemens de terre ou par l'explosion des volcans, comme le mont di Cenere, l'isle nouvelle près de Santorin, &c. mais il n'a pas pris garde que ces petites élévations formées par l'éruption d'un volcan ou par l'action d'un tremblement de terre, ne sont pas intérieurement composées de couches horizontales, comme le sont toutes les autres montagnes.[20]

In 1749, Buffon is thus convinced that the overall relief of the Earth's surface is to be explained by the action of bodies of water (sedimentation, erosion), an idea that for him provides a "key" with which to decipher the apparent disorder of the world. The system he proposes still takes account of certain incidental phenomena (i.e., catastrophes); he acknowledges their existence and is even prepared to advance some explanations for their activity. All in all, however, catastrophic phenomena have no *systematic* place in Buffon's theory of the Earth – they are exogenetic elements, accidents, which take no part in the order we discern in nature but, on the contrary, serve to disrupt it. Yet this is potentially very surprising. Why should Buffon take the trouble to point up the "énormes ravages produits par les tremblemens de terre"[21] if subsequently he is only to play them down? What should we make of all his thoroughly explicit assertions that – to reprise a passage cited earlier – the "mouvement naturel des eaux de la mer", "l'action des pluies, des gelées, des eaux courantes, des vents, des feux soûterrains, des tremblemens de terre &c."[22] have not insubstantially "contribué à changer la face du globe"?[23] Is there not a problem of coherence here?

My hypothesis is that it is only in order to guarantee the order which he believes he has discovered in nature that Buffon chooses to present the past as a succession of interminable cycles in the course of which the continuous action of barely perceptible causes (such as sedimentation from the ebb and flow of waters) has gradually shaped the Earth's surface. It is to preserve the continuity of his narrative (or rather its logical sequencing) that he refuses to recognize the importance of unpredictable and, in a word, exceptional events. In 1749 Buffon is *incapable*, so to speak, of according a significant role to

[20] Buffon, "Des volcans & des tremblemens de terre", *Histoire naturelle*, I, pp. 522-23.
[21] Buffon, *Histoire naturelle*, I, p. 522.
[22] Buffon, "Conclusion", *Histoire naturelle*, I, p. 611.
[23] Buffon, "Second discours", *Histoire naturelle*, I, pp. 109-10

natural catastrophes in his theory of the Earth; he does not possess a general structure – a conceptual framework or a metanarrative – that would allow him to conceptualize the impact of these unforeseeable, sudden catastrophic events while at the same time preserving the notion he has of the "order" of nature.

From the middle of the 1760s, however, Buffon started rethinking his theory of the Earth by imposing upon it another metanarrative – that of a systematic cooling – progressively remodelling it into the "system" that he would later set forth in the *Époques de la nature* (1778). This reconfiguration of his "geological" thought has its roots in the works of Dortous de Mairan "qui tendaient à prouver l'existence actuelle d'une chaleur propre du globe",[24] as well as in a series of experiments and calculations undertaken by Buffon in order to clarify a problem arising from his comparison of Newton's *Principia* with the same author's *Optics*. Without going into the details of this problem, suffice it to say that Buffon carried out a series of experiments on cooling which led him, quite naturally it seems, to indulge in some cosmogonical speculation. This manifests itself in his *Mémoire sur le refroidissement des planètes* which pre-dates the *Époques* by only a few years.

It is well known that in the *Époques* cooling processes constitute the "motor" of cosmogony. Thus during the first of Buffon's putative epochs, terrestrial matter (still in a liquid state following its ejection from the sun) takes the form of a spheroid; in the second epoch, it becomes solid from its surface through to its core. In the third epoch, the surface of the Earth is sufficiently cooled to accommodate the waters which had hitherto been suspended in the atmosphere; these then form a sea which covers the continents "jusqu'à quinze cents toises au-dessus du niveau de la mer actuelle".[25] The fourth epoch sees the collapse, under the weight of the waters, of subterranean caverns which had themselves been formed by the retreat of a part of the sea combined with the continued cooling of terrestrial matter. As Gabriel Gohau has noted, this is also the period when "les continents abritent

[24] Jacques Roger, "Buffon et l'introduction de l'histoire dans l'*Histoire naturelle*", in *Buffon 88*, ed. Jean Gayon (Paris: Vrin, 1992), p. 202. For more information, see Roger's "Introduction" to his critical edition of the *Époques de la nature* (Paris: Éditions du Museum, 1963), p. xxvii.

[25] Buffon, "Époques de la nature", in *Supplément à l'Histoire naturelle* (Paris: Imprimerie royale, 1778), V, p. 93.

des végétaux dont les débris ey sediments forment des charbons et des pyrites qui s'enflamment en s'unissant aux substances minerales sublimées par la grande chaleur de la terre, et produisent les éruptions volcaniques".[26] We will return to this last point later. During the fifth epoch, the progressive cooling of the Earth's surface steadily drives elephants and other so-called "animaux du midi"[27] towards lands strung along the equator which had hitherto been uninhabitable because they were too hot. These animals, born at the poles when the polar regions had the "même degré de chaleur dont jouissent aujourd'hui les terres méridionales",[28] follow the southerly movement of the temperature band in which they were born. As a result, by 1778 Buffon's theory of the Earth is wholly predicated on the analogy of a ball of metal heated up until it is white-hot then allowed to cool down. One of the most important consequences of the use of this new metanarrative is that natural catastrophes now fit into the temporal scheme of things. After all, do not the *Époques* expound an exact "history" of the Earth, the fourth epoch of which witnesses the first episodes of subsidence, the first earthquakes and volcanic eruptions – catastrophes that are caused by wholly natural and explicable mechanisms?

It is clear that between 1749 and 1778 the systemic status of catastrophe shifts: from being an exogenetic accident it becomes an endogenetic event, the product of a specific, identifiable dynamic, the seemingly necessary consequence of Buffon's privileged geo-physical hypothesis, that of an irreversible global cooling. In turn, this allows Buffon to re-evaluate the importance of natural catastrophes in the shaping of the Earth's surface. More than two decades after the publication of Lehmann's work, he comes to accept the existence of several "orders" of mountains: a primary order comprising anomalies, geological "blisters" forged by fire and transformed in cataclysmic events; and the remainder, formed by the action of water.

Towards an Aesthetics of Catastrophe

While the scientific – and epistemological – status of natural catastrophes changes over the long course of the *Histoire naturelle*'s publica-

---

[26] Gabriel Gohau, *Les sciences de la terre aux XVIIe et XVIIIe siècles. Naissance de la géologie* (Paris: Albin Michel, 1990), pp. 201-2.

[27] Buffon, "Époques de la nature", V, pp. 165, 170, 173, 188, 189.

[28] Buffon, "Époques de la nature", V, p. 165.

tion, other characteristics of catastrophe remain unchanged. Thus natural disasters invariably "prove" that nature is able to mobilize forces which defy all human sense of scale, reminding us that the world is fundamentally indifferent to the fate of the human race. Take, for example, those "feux intérieurs" whose explosions cause earthquakes and volcanic eruptions:

> rien n'est comparable à la force de ces matières enflammées & resserrées dans le sein de la terre, on a vû des villes entières englouties, des provinces bouleversées, des montagnes renversées par leur effort.[29]

These events partake of an order that we will never fully comprehend. Is it not the case, writes Buffon, that "nous ne pouvons juger que très-imparfaitement de la succession des révolutions naturelles; que nous jugeons encore moins de la suite des accidens, des changemens & des alterations; que le défaut des monumens historiques nous prive de la connoissance des faits"? He goes on:

> [I]l nous manque de l'expérience & du temps; nous ne faisons pas réflexion que ce temps qui nous manque, ne manque point à la Nature; nous voulons rapporter à l'instant de notre existence les siècles passez & les âges à venir, sans considérer que cet instant, la vie humaine, étendue même autant qu'elle peut l'être par l'histoire, n'est qu'un point dans la durée, un seul fait dans l'histoire des faits de Dieu.[30]

Besides, in the face of natural catastrophe man is not just a disembodied, rational observer; more often than not, he is a *victim* – a point which resonates greatly with Buffon. Hence it is no coincidence that, in the *Histoire naturelle*, the scientific study of catastrophic natural phenomena is accompanied by the progressive elaboration of a veritable poetics of natural violence. By way of illustration, here is Buffon's description of an active volcano:

> un volcan est un canon d'un volume immense, dont l'ouverture a souvent plus d'une demi-lieue; cette large bouche à feu vomit des torrens de fumée & de flammes, des fleuves de bitume, de soufre & de métal fondu, des nuées de cendres & de pierres, & quelquefois elle lance à plusieurs lieues de distance des masses de rochers énormes, & que toutes les forces humaines réunies ne pourroient pas mettre en mouvement; l'embrasement est si terrible, & la quantité

[29] Buffon, "Second discours", *Histoire naturelle*, I, p. 110.
[30] Buffon, "Conclusion", *Histoire naturelle*, I, pp. 611-12.

> des matières ardentes, fondues, calcinées, vitrifiées que la montagne rejette, est si abondante, qu'elles enterrent les villes, les forêts, couvrent les campagnes de cent & de deux cens pieds d'épaisseur, & forment quelquefois des collines & des montagnes qui ne sont que des monceaux de ces matières entassées.[31]

Obviously, such passages – and one could cite a number of them – do not consist of "purely" scientific prose. The vocabulary used here, the syntactical constructions, the scope of the rhetorical periods and the richness of figurative turns of phrase all hark back to the "grand style" of the Classical age. What we are dealing with here is a genuine "tableau poétique", an instance of hypotyposis, that is, a description characterized by a particularly vivid evocation of its object, what the Ancients called *enargeia.* According to Perrine Galand-Hallyn, this last term (synonymous with the Latin *euidentia* and *illustratio*) "relève de la rhétorique des affects";[32] it designates an "effet descriptif très particulier qui consiste à imposer à l'auditeur ou au lecteur l'image d'un objet ou d'un être absent".[33] This image is, so to speak, set "before the eyes" (*ante oculos*) of its audience. It is, of course, no coincidence that in the *Histoire naturelle* this rhetorical treatment is reserved precisely for the depiction of raging elemental forces and scenes of natural violence. By and large, this is in keeping with Galand-Hallyn's remark that "le catalogue des topiques propres à créer l'*enargeia*" centres precisely on "des joies et des craintes fondamentales de l'homme".[34]

Buffon's deployment of "poetic" or "enargetic" descriptions indicates that natural catastrophes, even if they are initially presented as objects of knowledge, can become objects of an aesthetic fascination. In other words, catastrophes can act as catalysts for the production of the "paradoxical" pleasure (paradoxical because mixed with terror) that the beholder feels when confronted with the spectacle of a vast and irregular natural vista or the natural forces unleashed in cataclysms on land or storms at sea, etc. This is, of course, what the eighteenth century calls "le sublime naturel"; and Yvon Le Scanff made an important point when he suggested that the sublime landscape finds its origin precisely in the ancient, rhetorical commonplace of the *locus*

[31] Buffon, *Histoire naturelle*, I, pp. 502-3.

[32] Perrine Galand-Hallyn, *Les yeux de l'éloquence. Poétiques humanistes de l'évidence* (Caen: Paradigme, 1995), p. 99.

[33] Galand-Hallyn, *Les yeux de l'éloquence*, p. 99.

[34] Galand-Hallyn, *Les yeux de l'éloquence*, p. 126

horribilis, which is opposed to the *locus amœnus* in exactly the same way that savage nature contrasts with a pleasure garden.[35]

We will have cause to return later to this notion of *locus horribilis*, since it provides a link between the aesthetics of natural catastrophe – clearly an aesthetics of the sublime – and Buffon's philosophical anthropology. But first let us make the important assertion that the two approaches to natural catastrophe (the scientific and the aesthetic) which we have so far considered are fundamentally incompatible. It is not that they suppose alternative visions of the same natural phenomena, it is rather that each seeks something completely *different* in nature itself. Therefore the incompatibility of scientific and aesthetics approaches to catastrophe does not reside in the play of emotions; the fundamental problem is that the force of the sublime resides in the unity and the uniqueness of the lived experience, whereas a natural disaster can only become an object of knowledge on the condition that it gives up its quality of uniqueness and fits into a series of comparable events whose differing causes can be analysed. The aesthete is captivated by a catastrophe which gives him the opportunity to lose himself in the totality of a unique occurrence; the scientist, on the other hand, is wary of this "observation-participation", and is only interested in phenomena insofar as their activity follows a recognizable pattern and their constituent elements are susceptible to analysis.

It is at this point that we should return to the notion of the *locus horribilis*. If this notion has such pertinence for our present study of the *Histoire naturelle* it is because with Buffon nature in the raw, as yet untamed by man, is invariably described as a "lieu d'horreur". It follows that the evocations of these terrible "sublime" cataclysms found in the *Histoire naturelle* correspond precisely to its author's conception of brute nature and serve above all to illustrate this. To reinforce this point, we might also look at the "tableaux" acting as décor in Buffon's descriptions of certain exotic fauna, for he considers that "ce n'est point en se promenant dans nos campagnes, ni même en parcourant toutes les terres du domaine de l'homme, que l'on peut connoître les grands effets des variétés de la Nature";[36] in order to see

[35] See Yvon Le Scanff, *Le paysage romantique et l'expérience du sublime* (Seyssel: Champ-Vallon, 2007).

[36] Buffon, "Le Kamichi", in *Histoire naturelle des oiseaux* (Paris: Imprimerie royale, 1780), VII, p. 336.

nature "telle qu'elle est"[37] we must leave behind those lands colonized by the human race and compare instead "les déserts avec les déserts".[38] Yet Jacques Roger is right to insist on the fact that "à mesure que l'*Histoire naturelle* s'éloigne de l'Europe et de ses paysages, marqués depuis des siècles par le travail de l'homme [...] la Nature devient farouche, hostile, hideuse".[39] This is a significant point: at a time when philosophy "choisit la voie de la précision et de la rigueur pour définir un idéal intellectuel" and when "la quête du bonheur se fait de plus en plus pressante", Buffon's scientific fascination for natural catastrophes that represent extremes of brute energy and disorder in nature is at the same time a horrified fascination betraying the presence of a genuine "métaphysique de l'inquiétude" in the beholder:

> [S]i, dans l'immédiat, les descriptions des ravages causés par les volcans provoquent le plaisir, elles permettent, à terme, de prendre conscience de la grandeur mais aussi de la fragilité tragique de l'homme: confronté à un univers physique hostile ou indifférent, celui-ci se sent en devenir l'étranger: les moments des grandes catastrophes [...] lui révèlent le possible néant d'une terre déserte.[40]

This means that in the *Époques de la nature* the sublime does not only constitute the culmination of a rhetorical or aesthetic project. Thanks to the force of its *enargeia*, Buffon is able to communicate what, on a grander philosophical level, is at stake beneath his scientific discourse.

The Anthropological Viewpoint

As we have seen, Buffon believes that catastrophes and other violent manifestations of natural power are only the most visible and spectacular proofs of nature's profound indifference towards mankind – an indifference which is present throughout "brute" nature and which can readily be interpreted as a mute hostility towards the human race. This

[37] Buffon, *Histoire naturelle des oiseaux*, VII, p. 336.
[38] Buffon, *Histoire naturelle des oiseaux*, VII, p. 337.
[39] Jacques Roger, *Buffon. Un philosophe dans le jardin du roi* (Paris: Fayard, 1989), p. 311.
[40] Dominique Peyrache-Leborgne, *La poétique du sublime de la fin des Lumières au romantisme* (Paris: Honoré Champion, 1997), p. 30.

is why Buffon states that man (and it is a question of *men* in Buffon) can expect nothing of nature unless he is able to impress his will upon it; he alone is responsible for the conditions of his existence and it is only through his own efforts that he can assure his survival. The whole of Buffon's philosophical anthropology is predicated upon this conviction. For we must not be mistaken: the texts Buffon devotes to mankind do not limit themselves to describing the "human condition" in all its diversity. They also – indeed above all – deliver a judgement on specifically human attitudes, customs, etc. Yet as Michèle Duchet has pointed out, the criterion for making this judgement never varies: it is "un certain rapport – puissance ou impuissance – de l'homme à la nature – éléments et espèces vivantes – qui définit l'état sauvage, l'état policé, l'état de civilisation".[41]

The importance of this binarism becomes clear when we consider that Buffon conceives of primitive man as a wretched being continually at risk of becoming the victim of natural forces that he barely understands and is incapable of controlling. Buffon thus opposes the proponents of a mythical Golden Age of humanity and refuses to accept that "dans le premier âge" man lived "sans inquiétude", "en paix avec lui-même [et] les animaux".[42] For him, it is absolutely no cause for regret that we have left behind our primitive state; the portrait he paints of primitive man at the start of the seventh "époque de la nature" is designed to drive the point home:

> Les premiers hommes témoins des mouvements convulsifs de la terre, encore récents & très fréquents, n'ayant que les montagnes pour asiles contre les inondations, chassés souvent de ces mêmes asiles par le feu des volcans, tremblants sur une terre qui trembloit sous leurs pieds, nus d'esprit & de corps, exposés aux injures de tous les éléments, victimes de la fureur des animaux féroces, dont ils ne pouvaient éviter de devenir la proie; tous également pénétrés du sentiment commun d'une terreur funeste, tous également pressés par la nécessité, n'ont-ils pas très promptement cherché à se réunir, d'abord pour se défendre par le nombre, ensuite pour s'aider & travailler de concert à se faire un domicile & des armes?[43]

Such passages demonstrate that humanity has not always been the "admirateur paisible de la nature" that we find described in other texts

[41] Michèle Duchet, *Anthropologie et histoire au siècle des Lumières* (Paris: Albin Michel, 1995), p. 246.
[42] Buffon, "Les animaux carnassiers", *Histoire naturelle*, VII, p. 26.
[43] Buffon, "Époques de la nature", V, p. 225.

by Buffon. Wild, primitive man is not the "maître du domaine de la terre"; he is but a terrorized victim, at the mercy of the "injures [des] éléments". Only society, "plénière & puissante, telle qu'elle existe parmi les peuples anciennement policés", is able to gain a proper hold over its natural environment.[44] It is equally no coincidence that Buffon offers up European societies as models of civilization, as the only societies in the eighteenth century possessing institutions capable of overseeing this sustained, rational drive to transform and domesticate nature. His "Première vue" of nature is very explicit in this regard, presenting man as a Prometheus using fire to create a "nature nouvelle":

> [M]ettons le feu à cette bourre superflue, à ces vieilles forêts déjà à demi consommées; achevons de détruire avec le fer ce que le feu n'aura pu consumer: bien-tôt au lieu du jonc, du nénuphar, dont le crapaud composoit son venin, nous verrons paroître la renoncule, le treffle, les herbes douces et salutaires; des troupeaux d'animaux bondissans fouleront cette terre jadis impraticable; ils y trouveront une subsistance abondante, une pâture toujours renaissante; ils se multiplieront pour se multiplier encore: servons-nous de ces nouveaux aides pour achever notre ouvrage; que le bœuf soumis au joug, emploie ses forces et le poids de sa masse à sillonner la terre, qu'elle rajeunisse par la culture; une Nature nouvelle va sortir de nos mains.[45]

Even so, Buffon is convinced that man's domination is always provisional; if he is not vigilant, nature quickly regains the ground she has temporarily ceded. Moreover, he believes that man's dominion – the dominion of fire – is slowly but inevitably heading towards its final catastrophe. Indeed, let us remind ourselves at this point that Buffon believes that our globe is constantly cooling down. As it cools, our world is progressively buried beneath endless sheets of ice, and the lifeforms it supports move inexorably towards extinction.

As we have seen, this idea subtends all the geological and biological reflections of the *Époques de la nature*. This vision persists in Buffon's later texts too, as in the following article of 1783 devoted to penguins and auks. Here the description of the "Terre des États" (New Zealand) and the Sandwich Islands provides him with the pretext for a brief but effective evocation of a world soon to be overwhelmed by

---

[44] Buffon, "Le castor", *Histoire naturelle*, VIII, p. 285.

[45] Buffon, "De la nature: première vue", *Histoire naturelle*, XII, p. xiii.

ice. In a sense, these polar regions only prefigure the fate awaiting the whole of the planet:

> [Ces] terres [sont] désolées, désertes, sans verdure, ensevelies sous une neige éternelle; nous [y] voyons [les pingouins], avec quelques pétrels, habiter ces plages devenues inaccessibles à toutes les autres espèces d'animaux, et où ces seuls oiseaux semblent réclamer contre la destruction et l'anéantissement, dans ces lieux où toute Nature vivante a déjà trouvé son tombeau. *Pars mundi damnata a rerum naturâ; æternâ mersa caligine* (Pline).[46]

It is here that the activist agenda that characterizes Buffon's philosophical anthropology finally shows itself for what it really is: a challenge issued to the human race. Whatever the brevity of human life, whatever the futility of our efforts in absolute terms, when faced with the ultimate catastrophe (a dead, frozen planet), mankind cannot *but* keep fighting to improve its lot on Earth.

Conclusion

These three means of apprehending natural catastrophe – one might say these three discrete discursive regimes of earth science, aesthetics and anthropology – co-exist then in the *Histoire naturelle*. But this coexistence does not in itself make them any more compatible with one another. We have seen why: catastrophe does not necessarily find its place in the order that humanity perceives in nature, for it is possible that events that appear perfectly normal – explicable, predictable, recurrent – on a global scale, appear as unforeseen and extraordinary on the human scale. How do we then reconcile the astonishment, fear and terror that a volcanic eruption causes to its victims with the attitude of the scientist for whom "tout cela n'est cependant que du bruit, du feu & de la fumée"?[47] As for the delight taken in a natural catastrophe when it is elevated to the status of a spectacle – that is quite another story again.

We must therefore conclude that catastrophe shatters the frame of reference that Buffon is in the habit of using. The "événe-

[46] Buffon, "Les pingouins et les manchots ou les oiseaux sans ailes", in *Histoire naturelle des oiseaux*, IX, pp. 377-79.
[47] Buffon, "Des volcans & des tremblemens de terre", *Histoire naturelle*, I, p. 503.

ment monstre" is too complex to allow itself to be comprehended in all its disquieting entirety by a single gaze; a gaze that would itself have to be at once that of the victim, the scientist and the spectator. But perhaps it is not necessary to desire at all costs that a coherence be found between these scientific, aesthetic and anthropological approaches. Imposing one on them would surely amount to a betrayal of the complexity of Buffon's thought. So perhaps we should content ourselves with remarking that the author of the *Histoire naturelle* quite simply accepts to pursue his research along divergent lines of enquiry, at the risk of losing the coherence of his vision of the world. Perhaps this is a trait of Buffon's modernity, a modernity that we share still.

Translated by David McCallam

When Geology Encounters a Real Catastrophe: From Theoretical Earthquakes to the Lisbon Disaster

Grégory Quenet

Abstract: *By the mid-eighteenth century earthquakes were already the object of scientific enquiry in France, albeit that they were often viewed with historical or geographical remoteness. The Lisbon earthquake of 1 November 1755 changed all of that. This chapter draws on national and local archives in order to demonstrate how the unmediated pan-European experience of the Lisbon shock enabled whole swathes of the population, hitherto untouched by the study of natural phenomena, to participate in one of the first truly popular scientific debates of the French Enlightenment. The continuing seismic activity after Lisbon, in France and elsewhere, also ensured that the Lisbon disaster threw up not only an array of new theories about earthquakes but also a more consistent rigorous protocol for observing and recording them. With Lisbon, as the chapter ably proves, natural catastrophe becomes an historical event.*

What must we do to combat earthquakes and to preserve humankind from the havoc they wreak? I would like to put forward here the suggestions of two respected scholars. The first proposes that:

> Dans chaque pais, des personnes passassent leur vie à examiner pendant tous les jours et les nuits avec une attention continuelle, sérieuse, méditée & réfléchie, les diférences et jusqu'aux plus petits changemens qui arrivent dans tous les élémens, et leurs diférentes parties, en suivant pas à pas, pour ainsi dire, leurs diférentes variations, dont ils feroient et conserveroient des nottes exactes; que ces observateurs après leur mort fussent remplacé par d'autres, qui auroient avec eux travaillés aux mêmes observations; et ainsi successivement jusqu'à ce qu'il

> fut arrivé plusieurs tremblemens de terre; pour lors raprochant ensemble toutes ces observations, l'on verroit si les circonstances qui ont précédé chaque tremblement sont les mêmes, et par là l'on pourroit découvrir quelques signes probables de l'aproche des tremblemens.[1]

The second proposition advances the idea of an alliance of peoples and nations joining forces to fight against a scourge that exceeds the capabilities of any single country: "puisse [*sic*] les souverains se liguer de concert pour détruire les fléaux multipliés qui semblent conjurés contre ce malheureux globe". You will not find any trace of these scholars in recent publications. The first is a modest participant in the essay competition organized in 1756 by the Académie de Rouen on the subject of the cause of earthquakes, one La Sablonnière le Jeune, a member of the Evreux religious chapter. The second is his more famous contemporary, the abbé Bertholon.

Just how and why did the scientific community devote its energies to explaining seismic activity and to combatting its effects after the famous Lisbon earthquake of 1 November 1755? The question may strike us as obvious, as much in light of the preoccupations of our own risk society as in view of Enlightenment concerns to overcome the misfortunes of our life in this world. But let us take a closer look. For, in fact, this is the first time that a real-life catastrophe mobilized the Republic of Letters with such urgency. Of course, historical earthquakes appear in earlier works, for example in Buffon's *Histoire et théorie de la terre*.[2] But these earthquakes are only ever cited as examples, most often in appendices and tables; they never constitute the basis for critical thought on the subject. Their ability to illustrate a theory prevents them from calling that same theory into question.[3] It took the Lisbon disaster to throw the contrast between theoretical and real earthquakes starkly into relief.

[1] Bibliothèque Municipale de Rouen, fonds de l'Académie, C 20: Concours de 1756 sur les tremblements de terre, mémoire n° 9 (fol. 2).

[2] The reader is referred here to the previous article by Benoît de Baere.

[3] To my knowledge, historians of geology do not use the term "risk", as if the history of their texts had nothing to do with the objective prediction of future catastrophes.

A Classical Topos of the Scientific Imagination: The Destruction of a City

What was it then that really grabbed the scientific imagination? Chronologically, the destruction of Lisbon came first and acted as the trigger for what followed. By 1755 the Portuguese capital had been in decline for some time and no longer occupied an important place on the European stage. It nonetheless retained a certain prestige, since gold and riches from the Americas still fired the contemporary imagination. Stories of the earthquake all pick up on this image:

> Lisbonne étoit l'une des plus grandes villes de l'Europe, bâti sur un terrain fort élevé & montagneux, formant un Amphithéâtre qui procuroit en perspective un très beau coup d'œil. L'on y remarquoit de très beaux édifices, parmi lesquels le Palais du Roi étoit véritablement digne par sa magnificence de loger un souverain [...] On y voyoit de très beaux couvents, grands & spacieux, peuplés d'une multitude de religieux & religieuses, & en si grand nombre que l'on en pouvoit compter près de deux cens dans une seule Communauté. Les ornements des Autels étoient la plupart enrichis d'or, d'argent, de pierres précieuses, de perles fines, & de brillants en quantité.[4]

More generally, however, the razing of the city to the ground was a commonplace of the scientific imagination of the time. In classical culture, it was conceived of as a complete reversal of fortunes, epitomized by the disaster in Asia which completely obliterated twelve cities in a single night at the beginning of Tiberius's reign. This incident is reported by Tacitus, Seneca, Strabo, and is later reprised by Coeffeteau in his *Histoire romaine*, by Diderot and d'Alembert in the *Encyclopédie*, and by a host of other eighteenth-century writings on earthquakes.[5] The discoveries of Herculanum (1719) and Pompeii (1748) only reinforced this image.[6] In fact, it becomes the prevailing

---

[4] G. Rapin, *Le Tableau des calamités, ou description exacte et fidèle de l'extinction de Lisbonne*: ([n. p.], 1756), pp. 3-4.

[5] Seneca, *Naturales quaestiones*, Book VI, 1; Tacitus, *Annales*, Book II, 47; Strabo, *Géographie*, Book XII, 8, 16-18; Nicolas Coeffeteau, *Histoire romaine* (Paris: 1646), p. 275; *Encyclopédie ou Dictionnaire raisonné des sciences, des arts et des métiers*, ed. Denis Diderot et Jean Le Rond d'Alembert, 35 vols (Paris; Neuchâtel: 1751-1780), XVI, pp. 580-83.

[6] Antoine-Léonard Thomas, *Mémoire sur la cause des tremblemens de terre* (Paris: 1758), p. 84.

representation of earthquakes in scientific writings before Lisbon. As Buffon declares in his *Histoire et théorie de la terre*, "rien n'est comparable à la force de ces matières enflammées et resserrées dans le sein de la terre, on a vu des villes entières englouties, des provinces bouleversées, des montagnes renversées par leur effort".[7] The force of the image does not only come from its occurrence in classical literature. The Ancients remained an important source of information about earthquakes in earlier periods. Thus when the Geographer Royal, Philippe Buache, compiled a dossier on earthquakes, he drew up a chronological table dating back more than a thousand years before the Christian era, indicating "[ce] dont les historiens nous ont parlé en faisant le récit de la destruction des villes, du bouleversement des terres et des éruptions des volcans".[8] The same image appears in dictionaries of the time, as in Furetière who wrote in 1690 that "les tremblements de terre renversent les villes & les montagnes, changent le cours des rivières, &c. L'Italie et les pays orientaux sont sujets aux tremblements de terre".[9]

The minutes of the Académie royale des sciences provide us with yet another means of perpetuating this image of earthquakes: the information networks among academicians. In fact, up to 1744, the academicians were particularly well informed about regions outside France, even outside Europe. References to seismic activity in European countries were limited to citing Portugal in 1723 and Italy in 1703, 1704 (two references) and 1742. In contrast, Jesuit missionaries reported the occurrence of earthquakes in China and in Santo Domingo to the Académie; this was the result of the learned society reaching an agreement with Jesuits travelling in the Far East so that the missionaries would provide them with scientific data, a collaboration which in 1692 would result in the Académie's *Observations Physiques et mathématiques envoyées des Indes et de la Chine*.[10] Diplomats also contributed to data collection. For instance, the former

[7] Georges Louis Leclerc, comte de Buffon, *Histoire et théorie de la terre* (Paris: Imprimerie royale, 1749), I, p. 110.

[8] Paris, BnF, ms. n.a.fr. 20236 et 20237: Philippe Buache, *Mémoire explicatif de la carte des tremblements de terre dressée par Buache*.

[9] "Tremblement", in Antoine Furetière, *Dictionnaire universel*, 3 vols (La Haye; Rotterdam: 1690), III, (not paginated).

[10] Roger Hahn, *The Anatomy of a Scientific Institution: The Paris Academy of Sciences, 1666-1803* (Berkeley: University of California Press, 1971); French trans., (Paris:Yverdon, 1993), p. 24.

ambassador to Constantinople reported the tremors felt on the island of Santorini in 1712, and the French Consul in Lisbon reported those in the Azores and in the Algarve in 1722 and 1723.[11] Finally, the contribution of travellers was to remain paramount until the end of the eighteenth century, supplemented as it was by accounts from merchants and soldiers, and especially from naturalists and scholars, such as Père Feuillée (astronomer, explorer, naturalist), Jean-Baptiste Lignon le jeune (botanist to the king, in Guadeloupe), Louis Godet (astronomer, and sometime resident of Lima, Peru), and Martin Folkes (physician and archaeologist) – the list goes on.

This school of geography necessarily privileged destructive earthquakes and the terrifying descriptions of them. Moreover, for Parisian readers, the remoteness of these phenomena could only enhance their extraordinary character. The acme comes with upheavals of the Earth's crust, with mountains bunched together or razed to the ground, with villages swallowed up, waters run dry, or stretches of land torn apart. Read, for instance, the tale of the "furieux tremblement de terre arrivé dans le Chensi" in China in June 1718:

> Au nord de la ville de Tong Ouei la terre s'ouvrit, les montagnes tombèrent, et en tombant roulèrent dans la ville par le coin du nord, et passèrent vers le midy, de manière qu'en un clein d'oeuil toute la ville fut engloutie, et la plaine s'enfla et s'éleva à la hauteur de plus de six toises, sans qu'il soit demeuré une seule maison sur pied; les greniers publics, l'argent du trésor, les prisons et les prisonniers tout fut enseveli sous terre, et de dix personnes à peine s'en pût-il sauver deux ou trois. De toute la famille du gouverneur nommé Hoang, il s'est sauvé seul avec son fils et un valet.
>
> A Tsing Nig Tchin depuis trois heures du matin jusqu'à onze la terre trembla, les édifices publics, et les murs du côté du midy furent abatus. Le mont Outai tomba plus qu'à la moitié au midy ; il y eut une infinité d'hommes et d'animaux tuez ou blessez.[12]

Devastating events fuelled the debate about the effects of earthquakes, a classic question that Descartes had revisited in the seventeenth century. According to his enquiries, earthquakes could not shake the foundations of the globe because they were only the result of superfi-

[11] Paris, Académie royale des Sciences, Registre des Procès-verbaux des séances, 29 mars 1713, vol. 32, fol. 111v-114 ; *Ibid.*, 2 septembre 1722, vol. 41, fol. 264 ; *Ibid.*, 1723, vol. 42, fol. 39-40v.

[12] Académie royale des Sciences, Registre des Procès-verbaux, 29 novembre 1719, vol. 38, fol. 288v-289v.

cial causes, namely the combustion of pockets of gas trapped in cracks in the Earth's crust; and he compared this process to the thick smoke released by candlesticks when a flame is brought near the wax.[13]

Returning to the Lisbon disaster, the shock it caused had less to do with the levelling of a city *per se* than it did with the destruction of what Lisbon in particular represented for contemporaries. Seismic activity was regular in the Lisbon area, the city having suffered seven earthquakes in the fourteenth century, seven again in the sixteenth century, three of which were very violent (1,500 houses destroyed in 1531; 2,000 dead in 1551; three streets swallowed up in 1597), a further three in the seventeenth century, and two more in the first half of the eighteenth century (in 1724 and 1750).[14] Yet these events were little known outside the Iberian peninsula, and Portugal was not regarded as a particularly seismic region. In France, news-sheets, periodicals and literary texts prior to 1755 do not mention Lisbon or Portugal as seismic zones, whereas earthquakes occurring in Italy and the Eastern Mediterranean are frequently cited.[15] The truly extraordinary nature of the disaster of 1 November 1755 only seems greater as a result of this. In the stories published about it, the seismic violence overshadows the tsunami and the fire which nonetheless caused the majority of the devastation.[16] Other earthquakes in the early modern period attracted scientific comment, such as those in China in 1699, London in 1750, and Lima in 1751; but none allowed for the mass participation in the event which characterized the Lisbon disaster and which made it a founding date of European consciousness.

An Unprecedented Participation in the Event

Contemporaries were as much struck by the tremors that affected Europe for several months afterwards as they were by the Lisbon earthquake itself. The effects of the seism of 1 November 1755 were

---

[13] Cited in Gabriel Gohau, *Les sciences de la terre aux XVIIe et XVIIIe siècles. Naissance de la géologie* (Paris: Albin Michel, 1990), pp. 80-81.

[14] José Augusto Franca, *Une Ville des Lumières, la Lisbonne de Pombal* (Paris: Fondation Calouste Gulbenkian, 1988), p. 52.

[15] The remark does not really apply after 1755 when contemporaries trawled through history to find traces of earlier earthquakes at Lisbon.

[16] Augusto Franca, *Une Ville des Lumières*, pp. 58-59.

felt in northern Europe, northern Italy, Catalonia, the south of France, Switzerland, Bohemia, the Azores, along the coast of Brazil, in the West Indies, from Iceland to Morocco, and from the German provinces to Boston, America.[17] In the months following the earthquake, the exceptional number of shocks allow us to speak of the "crise séismique de 1755-1762 en Europe du Nord-Ouest".[18] France was hit by at least forty-four earthquakes between 1 November 1755 and the end of 1756, and by eighty-seven in total up to 1762. All the inhabitants of northern Europe at one time or another felt the earth shake in the months following the Lisbon earthquake. A single German source from the Eifel region counted eighty-eight tremors between December 1755 and March 1757. The rector of the Jesuit college at Brigue recorded 135 tremors from 9 December 1755 to 26 February 1756, leaving only twenty-six days free of seismic activity out of a total of eighty, with 9 December 1755 proving the most active, punctuated by a tremor almost every half-hour.[19]

All the stories, public or private, emphasize the exceptional scale and frequency of the tremors through 1755 and 1756, thereby linking them to the events in Portugal. A good number of French priests note in the parish registers the coincidence of an earth tremor felt in their village with the shock visited on Lisbon.[20] These, then, are Europe-wide events which galvanize into action the scholars of the Académie des sciences in Paris, although it is no longer solely about Lisbon. Only two of the ninety-two papers on earthquakes presented in the Académie between November 1755 and the end of 1756 relate

[17] Académie royale des Sciences, Registre des Procès-verbaux des séances, 31 janvier 1756, vol. 75, fol. 55.

[18] Pierre Alexandre et Jean Vogt, "La crise séismique de 1755-1762 en Europe du Nord-ouest. Les secousses des 26 et 27.12.1755: recensement des matériaux", in *Materials of the CEC project "Review of Historical Seismicity in Europe"*, ed. Paola Albini et Andrea Moroni (Milan: 1994), vol. 2, pp. 37-76.

[19] Alexandre and Vogt, "La crise séismique", p. 37; Frédéric Montandon, "Les séismes de forte intensité en Suisse", *Revue pour l'étude des calamités* (1942-1943), 5-6, pp. 9-10.

[20] See for example the Loire-Atlantique, archives communales de Soudan, registre paroissial manuscrit (1755); Loire-Atlantique, archives communales de Saint Sulpice d'Auvergné, registre paroissial manuscrit (1755); Haute-Savoie, archives communales de Cernex, registre paroissial manuscrit (1755); Oise, archives communales de Hedencourt, registre paroissial manuscrit (1755).

to Portugal.[21] Again, of the ninety-six sets of correspondence gathered together by the geographer, Philippe Buache, from his network of academic contacts, only one concerned Lisbon, and this is dated 1757.[22] From a seismological point of view, the majority of these earthquakes were neither aftershocks nor effects of the tremors of 1 November 1755. Some of them were very powerful, such as those of 9 December 1755 in the Valais region of Switzerland (intensity VIII-IX at Brigue), 27 December 1755 in the Rhineland (intensity VII at Stolberg), and 18 February 1756 in the same region (intensity VIII at Stolberg).[23]

Light earth tremors, which would hitherto have gone unnoticed, were recorded and commented on because they were thought to relate to the earthquake of 1 November. More attentive to seismic phenomena, recording them with more sophistication and relating them to one another with greater consistency, observers at the time convinced themselves that the shocks were becoming more frequent after 1755. Jean-Baptiste Robinet was only repeating a commonplace of the time when he wrote: "jamais les secousses de tremblement de terre ne furent ni si étendues ni si fréquentes, que depuis quelques années".[24] By means of a social dynamic which magnified the perception of risk, this natural hazard came to the forefront of contemporary preoccupations.[25] Several sources point to the existence of an oral tradition in France perpetuating the memory of the Lisbon earthquake. The finest scientific instance of this is arguably provided by Philippe Buache himself when Paris was hit by a tremor on 18 February 1756.

---

[21] Académie royale des Sciences, Registre des Procès-verbaux des séances, 1755-1756, vol. 74-75.

[22] Bibliothèque Nationale de France, ms. n.a.fr. 20236 et 20237.

[23] *Les Tremblements de terre en France*, ed. Jérôme Lambert (Orléans: BRGM, 1997), p. 191. The MSK scale is a measure of macroseismic intensity, that is, it registers the intensity of a shock on the basis of its effects on buildings and people. It is used in France and in the majority of European countries because it is suited to regions of low seismicity. It comprises twelve degrees, but should not be confused with the more famous Richter scale which measures the magnitude of an earthquake, quantifying the power of a shock as represented by the energy radiating from the epicentre in the form of seismic waves.

[24] Jean-Baptiste Robinet, *De la nature* (Amsterdam: 1761), p. 64.

[25] William J. Burns, Jeanne X. Kasperson, Roger Kasperson , Ortwin Renn, Paul Slovic, "The social amplification of risk: theoretical foundations and empirical applications", *Journal of social issues*, 48 (1992), pp.137-60.

After having related numerous accounts by Parisians to his fellow academicians, he noted in his handwritten papers that:

> A l'égard de l'isle du Palais où je demeure dans la partie occidentale qui est formée de terres rapportées, je ne me suis aperçu d'aucun mouvement quoique je fusse ce jour, dès 5 h du matin, à travailler avec beaucoup de tranquillité à l'arrangement d'une table, par ordre alphabétique, des lieux où les tremblements se sont fait sentir depuis quelques mois en Europe et ailleurs. Mais j'ai appris que dans la partie orientale qui est le terrain solide de l'isle, c'est-à-dire dans l'hôtel des Ursins, on s'étoit aperçu d'un mouvement qui avoit effrayé des personnes qui travaillaient dans leur cabinet.[26]

The Lisbon earthquake would not have had such an impact if contemporaries had not felt the same shocks at thousands of kilometres' distance yet at precisely the same moment. In 1755-1756 nature forged in a quite singular way a connection between a distant event and its Europe-wide audience. This connection enabled a remote yet almost instantaneous participation in the event, a characteristic, according to Pierre Nora, more commonly associated with the modern media.[27] Nora effectively situates the emergence of the media event at the end of the nineteenth century when progress in communications, and the growing integration of distant parts of the world via colonialization, world war, revolution and economic interdependence, allow for the democratization of, and mass participation in, events. The Lisbon earthquake therefore realized an unprecedented unification of European society, something possibly without equal before the outbreak of the French Revolution.[28] At a time of basic – albeit ever-improving – information systems, these seismic shocks suddenly collapsed geographical distance between the remotest parts of the continent, affording a totally new involvement in the event. Consequently, there was a shift in the perception of nature, a new sensibility to tremors, something clearly manifest in the numerous journals which re-

[26] Académie royale des Sciences, Registre des Procès-verbaux des séances, 21 février 1756, vol. 75, fol. 99.

[27] Pierre Nora, "Le retour de l'événement", in *Faire de l'histoire, Nouveaux problèmes*, ed. Jacques Le Goff and Pierre Nora, 3 vols (Paris: Gallimard, 1974), I, pp. 285-308.

[28] Ana Cristina Bartolomeo de Araujo, "1755: l'Europe tremble à Lisbonne", in *L'Esprit de l'Europe, dates et lieux*, ed. Antoine Compagnon and Jacques Seebacher, 3 vols (Paris: Flammarion, 1993), I, p. 126.

corded ten to twenty earthquakes in the course of the year after Lisbon.[29] Each of these phenomena is superadded to the Lisbon earthquake to form an "événement-monstre", an event of freakishly colossal proportions, while also convincing contemporaries of the increasing frequency of seismic activity.

The increase in incidence of earthquakes gave rise to a new literary genre: telluric tables or journals. In 1756 Anne Amable Augier du Fot published a *Journal historique, géographique et physique de tous les tremblements de terre et autres événements arrivés dans l'Univers pendant les années 1755 & 1756.*[30] The following year Laurent-Etienne Rondet accompanied his *Supplément aux réflexions sur le désastre de Lisbonne* with a *Journal des phénomènes et autres événements remarquables arrivés depuis le 1er novembre 1757.*[31] Mention must also be made of Elie Bertrand's *Mémoires historiques et physiques sur les tremblements de terre* as well as the tables drawn up by Philippe Buache.[32] As a result of the events in 1755-1756, Gueneau de Montbeillard started to draft a chronological list of earthquakes often considered to be one of the first seismological catalogues.[33] Moreover, this genre of writing is a European phenomenon, represented by German and Dutch works as well.[34] Histories of

---

[29] *Jacquemin, cultivateur à Aische en Refail: annotations pour les années 1755-1760*, ed. E. Verhelst, "Etude de géographie locale: Aische en Refail", *Bulletin de la Société royale de Belgique*, 19 (1895), pp. 548-49.

[30] Anne Amable Augier du Fot, *Journal historique, géographique et physique de tous les tremblements de terre et autres événements arrivés dans l'Univers pendant les années 1755 & 1756* (n.p.: 1756).

[31] Laurent-Etienne Rondet, *Supplément aux réflexions sur le désastre de Lisbonne. Avec un journal des phénomènes, et autres événements remarquables arrivés depuis le 1er novembre 1755 et des remarques sur la plaie des sauterelles annoncée par saint Jean* (n.p.: 1757).

[32] Elie Bertrand, *Mémoires historiques et physiques sur les tremblemens de terre* (La Haye: 1757). Bibliothèque Nationale de France, MS Nouvelles acquisitions françaises 20236 et 20237, "Tables alphabétiques des lieux où l'on a ressenti des tremblements et leur Supplément par Philippe Buache".

[33] Philippe Gueneau de Montbeillard, "Liste chronologique des éruptions de volcans, des tremblements de terre, de quelques faits météorologiques, des comètes, des maladies pestilentielles, des éclipses les plus remarquables jusqu'en 1760", in *Collection académique composée des mémoires, actes ou journaux des plus célèbres académies et sociétés littéraires de l'Europe* (Paris: 1761), VI, pp. 450-700.

[34] *Schouwtoneel der akelige en deerlyke verwoestingen, rampen, ongevallen en zonderlinge gebeurtenissen, Sedert den eersten November 1755 zo in Portugal, Spanje, Vrankryk, Italie, Zwitzerland, Duitschland, het Noorden, Engeland en de*

earthquakes were not, of course, a novelty in themselves; they had been widespread in the sixteenth century, and remained common in the early eighteenth century, as is illustrated by the work of Johann Gottlob Krüger.[35] By 1756 the novelty lay not in seeking seismic phenomena among the Ancients or in sacred texts, but in the present and the very recent past on the basis of information gleaned from gazettes and collections of correspondence. This wealth of contemporary information fuelled the philosophical debate about the nature of evil, but very often went far beyond this too.

When Scientists Interrogate the Earth

From 1756 onwards earthquakes provoked fierce debate in the physical sciences, a debate rivalling in intensity the metaphysical polemics over the presence of evil in the world. The physical science debate constitutes two-thirds of all the articles published on earthquakes in the *Mercure de France*, a third of all those appearing in the *Journal de Trévoux*, half of those in the *Journal des savants*, and over half again in the *Journal encyclopédique*, with the rest comprising literary, religious or philosophical reflections.[36] It is not by chance that Kant's dissertation of 1756 was devoted to seismological theory rather than the workings of Providence.[37] Investigations focussed above all on the scale and the propagation of tremors, as these surpassed all known

Nederlanden, als buiten Europa door de Aardbevingen, waterberoeringen Overstromingen en zeldzame Luchtverschynsels verwekt en voorgevallen (Utrecht: 1756); Johann-Friedrich Seyfart, *Allgemeine geschichte der erdbeben* (Frankfurt; Leipzig: 1756).

[35] *Discours des causes et effects admirables des tremblemens de terre, contenant plusieurs raisons & opinions des philosophes. Avec un brief recueil des plus remarquables tremblemens depuis la création du monde jusques à present, extraict des plus signalez historiens par V.A.D.L.C.* (Paris: 1580); Johann Gottlob Krüger, *Histoire des anciennes révolutions du globe terrestre avec une relation chronologique et historique des tremblemens de terre arrivés sur notre globe depuis le commencement de l'Ere chrétienne jusqu'à présent*, trad. M. F. A. Deslandes (Amsterdam; Paris: 1752).

[36] Grégory Quenet, *Les tremblements de terre aux XVIIe et XVIIIe siècles. La Naissance d'un risque* (Seyssel: Champ-Vallon, 2005), pp. 74-89.

[37] Immanuel Kant, "Histoire et description des plus remarquables événements relatifs au tremblement de terre qui a secoué une grande partie de la terre à la fin de l'année 1755", trans. Jean-Paul Poirier, *Cahiers philosophiques*, (mars 1999), pp. 85-121.

phenomena hitherto and turned notions of space on their head. How could an event occurring in Lisbon be felt at the very same instant thousands of miles from there? Having previously sought to address the role of earthquakes in Earth history, existing models of thought were unable to answer the question.[38] This is emphasized by the editor of one collection of seismological reflections published in London in 1757: "the effects of the earthquake [...] were distributed over very nearly four millions of square English miles of the earth's surface: a most astonishing space! and greatly surpassing any thing, of this kind, ever recorded in history".[39] The *Encyclopédie* revisited the same question, since "un des phénomènes les plus étranges des tremblemens de terre, c'est leur propagation, c'est-à-dire la manière dont ils se communiquent à des distances souvent prodigieuses, en un espace de tems très-court".[40]

In the months following the spate of Europe-wide tremors, dozens of new theories were elaborated and published. They are extremely diverse in character, as is illustrated by the fact that of the ten extant essays submitted to the Académie de Rouen's essay competition on earthquakes in 1756 no two essays propose the same explanatory model. These texts are in no way content simply to reprise the Ancients' views on earthquakes. Of course, they cite them and borrow widely from them; but borrowings and citations relate to specific, quite concrete, points. No treatise of the 1750s adopts the Ancients' general cosmology, nor do they accept in its entirety Aristotle's theory of meteors, which was nonetheless ubiquitous in the first half of the seventeenth century. Earthquakes were still a poorly constituted object of knowledge, since they did not belong to any single branch of science, and since it was impossible to prove the explanations put forward for them. As Kant writes:

> Nous connaissons la surface de la terre à peu près complètement en ce qui concerne son étendue. Mais nous avons aussi sous nos pieds un monde, avec lequel, encore à notre époque, nous sommes très peu familiers. [...] La plus grande profondeur à

[38] Grégory Quenet, *Les tremblements de terre*, pp. 305-54.

[39] *The History and philosophy of earthquakes, collected from the best writers on the subject by a member of the Royal Academy of Berlin with a particular account of the great one of November, the 1st 1755 in various parts of the globe* (London: 1757), p. 5.

[40] "Tremblemens de terre", in *Encyclopédie ou Dictionnaire raisonné*, 35 vols (1751-1780), XVI, pp. 580-83.

> laquelle les hommes soient parvenus, à partir des endroits les plus élevés de la terre ferme, ne dépasse pas toujours cinq cents brasses, c'est-à-dire même pas la six millième partie de la distance au centre de la terre.[41]

Or, as a less famous contemporary expressed it:

> Pour acquérir des connaissances certaines sur les circonstances qui précèdent les tremblemens de terre, et découvrir les signes de leur approche il faudroit que les tremblements fussent ou à peu près périodiques, ou très fréquens, comme la pluye, le vent, la grêle, mais ces tristes événements ont été si rares jusqu'à présent, que qui que ce soit ne s'est appliqué à faire de sérieuses observations.[42]

The shocks of the 1750s elicited quite an array of novel scientific approaches, with each being put to the test in order to determine the best theoretical model. As the scientific domain was neither fully professionalized nor totally discrete, the public nature of the debate legitimized each and every intervention. Examples of this include the mineralogical theories on underground conflagration; variations on the chemical theme of fermentation and dilation of air pockets; Philippe Buache's investigations into the propagation of tremors via mountain chains; mechanical models of pulse transmission; and not forgetting disquisitions citing celestial motion, the circulation of the elements and the air, subterranean vaults, phlogiston theory, etc.[43]

Without a doubt the most successful and popular of these theories invoked electricity as the cause and means of propagation of earthquakes. With just such a theory, Isnard from Grasse won the Académie de Rouen's essay competition, beating Antoine-Léonard Thomas's hypothesis of fermentation-dilation into second place.[44] These electricity-inspired explanations were themselves however very varied, as for some writers earthquakes resulted from the contact of an electrical body with one that is not; for others, earthquakes were produced by an electrical pulse of mysterious origin; while yet others contended that an electrical charge ignited sizeable sulphur deposits

---

[41] Immanuel Kant, "Histoire et description", p. 86.

[42] Bibliothèque municipale de Rouen, fonds de l'Académie, MS C 20, "Concours de 1756 sur les tremblements de terre, mémoire n° 10 par la Sablonnière le Jeune du chapitre d'Evreux".

[43] Grégory Quenet, *Les tremblements de terre*, pp. 357-95.

[44] Isnard, *Mémoires sur les tremblemens de terre, qui a remporté le prix de physique au jugement de l'Académie des sciences, belles-lettres et arts de Rouen, le 3 août 1757* (Paris: 1758); Thomas, *Mémoire sur la cause des tremblemens de terre*, p. 84.

buried deep in the Earth's crust.[45] Yet for all these differences, the mechanism for propagating earth tremors remained the same, as the "fluide électrique" was transmitted instantaneously through all conducting bodies, whether they be mountains, veins of sulphur, or other substances. For Isnard, the electrical "principle" is the only one capable of explaining all the particular effects of earthquakes: electricity penetrates objects without losing anything of its power, it strikes matter both externally and internally, it produces detonations and showers of sparks, it passes through the hardest bodies as lightning does, and is luminescent. The most compelling argument for contemporaries remained electricity's ability to travel enormous distances in a trice just as earth tremors did.[46] Isnard cites Louis-Guillaume Le Monnier who in 1746 was the first to estimate the speed of electricity: detecting no time delay between seeing lightning flash and feeling it strike, he deduced that the electrical charge travelled at thirty times the speed of sound. Le Monnier also noted that an electrical current crossed an expanse of water, such as the pond in the Tuileries gardens, without losing any of its power. Even more spectacular experiments were carried out in this vein by Jean Jallabert around Lake Geneva and by William Watson and other members of the Royal Society in London in 1748 when they would send currents across rivers by means of two iron bars plunged simultaneously into the water.[47]

[45] "Les tremblemens de terre attribués à l'électricité", *Journal encyclopédique* (1er mai 1756), III, pp. 3-18; "Essai sur les tremblemens de terre", *Mercure de France*, (mai 1756), pp. 93-113; "Réflexions sur les causes des tremblemens de terre", *Journal de Trévoux*, (décembre 1756), pp. 3012-16.

[46] Isnard, *Mémoires sur les tremblemens de terre*, p. 27.

[47] John Lewis Heilbron, *Electricity in the 17th and 18th centuries: a study of early modern physics* (Berkeley; London: University of California Press, 1979), p. 320.

A New Way of Looking at Nature

The problem posed by earthquakes was not only theoretical but also practical: the tremors of 1755-1756 necessitated new methods of investigation. In order properly to study the scale and distribution of this phenomenon, a greatly increased number of observations were needed, as it was no longer viable to limit data to a single account specific to its locality. Each scientist had therefore to collate information by calling on his network of correspondents. In Paris this meant putting together an overview drawn from various localities across the country – Aix, Toulouse, Sedan, Beaune, and many more besides. This empirical history in fact started in the 1740s when scientists first encountered earthquakes in France. These low-to-moderate intensity phenomena were more numerous than had previously been thought, forcing scientists to construct new grids of reference and appropriate procedures for observing them.

The protocol for observing earthquakes thus appeared in the minutes of the Académie des sciences in the decade before the Lisbon disaster. Responding to a request from René-Antoine Réaumur, Chomel de Bressieu provided an account of the earthquake which hit Annonay in 1740, giving the time of the shock, its duration in seconds, the area affected, the effects felt and further comments (for example, the differing sensations according to which floor of a building one was on).[48] Similarly, in 1750, in the Académie de Toulouse the astronomer Antoine Darquier read letters sent to the Académie by a correspondent in Tarbes concerning the earthquakes which had recently struck the Bigorre region.[49] With the increase in seismic research and observations in 1755 and 1756, the criteria to be used were honed and fixed. Scientific accounts came to resemble the letter written from Geneva by Jean Jallabert in response to questions from Jean-Jacques Dortous de Mairan about the earthquake of 9 December:

> C'étoit 2 heures 23' après midi. La plupart n'ont senti que deux secousses distantes l'une de l'autre d'environ 30''. Quelques personnes ont cru en avoir remarqué 3. Je les jugeai sur la direction des oscillations de quelques corps suspendus du sud-est au

[48] Académie royale des Sciences, Registre des Procès-verbaux des séances, 23 mars 1740, vol. 59, fol. 58.

[49] "Sur un tremblement de terre, & sur des effets singuliers de la foudre", in *Histoire et Mémoires de l'Académie des sciences, inscriptions et belles-lettres de Toulouse* (Toulouse: 1784), II, pp. 15-19.

> nord-ouest. Le baromètre était à 26 pouces 6 lignes 1/2 le thermomètre a I d ½ au-dessus de 0. Le vent très foible à l'ouest, les jours précédents avoient été les plus froids de l'année avec le vent au nord. A 7 heures du matin, j'observai le 6 le thermomètre de M. de Réaumur à 6 ½; le 7 à 7 ¼, le 8 à 7, et le 9 à 0. [...] Tout l'effet du tremblement s'est borné à renverser quelques outils dans des cabinets d'horlogers logés au haut des maisons, et à faire sonner quelques cloches.[50]

To the initial reporting framework were added the lapse of time between shocks, their direction, the weather conditions (air pressure, temperature, wind direction, etc.), specifying where necessary any coincidence with other unusual phenomena (storms, comets, fogs). With a greater or lesser degree of rigour and exhaustiveness, the reports on seismic activity sent to the Académie des sciences followed this pattern.

Knowing the time at which an earthquake struck was fundamental. A simultaneous shock occurring at two different points allowed scientists to conclude that they were dealing with the same phenomenon and a single cause. If a slight chronological discrepancy existed between the shocks, this focussed investigations on the physical mechanics of the earthquake: how did a shock occurring at a central point spread? What was the speed of its propagation? Such questions demanded an increased level of precision in order to corroborate proofs, calculate velocities and establish the physical geography of an area. Scientific accounts of the time paid close attention to determining the direction of land movements where these occurred so as to arrive at a more general understanding of the phenomenon. Directions of land movements indicated the epicentre of earthquakes and synthesizing this sort of information enabled scientists to link up spatially dispersed events. Measuring the duration of shocks provided yet another element of comparison as well as suggesting further trails of enquiry for geophysical theory. It was (and is still) an implicit means of assessing the intensity of the shock. Empirically, contemporaries believed that the duration of the tremor determined, among other things, the extent of the material damage sustained. Comparing the earthquake that hit Aix on 3 July 1756 with the firsthand accounts of the Lisbon disaster of 1 November 1755, one witness of the former explained that "les secousses furent moins violentes [à Lisbonne] que celles que nous avons ressentis ici, mais qu'à la vérité elles furent

[50] Académie royale des Sciences, Registre des Procès-verbaux des séances, 31 janvier 1756, vol. 75, fol. 55.

beaucoup plus longues. Si celle-ci avoit duré quelques minutes toute la ville auroit été abîmée".[51] This desire for accuracy, however, did not prevent certain witnesses from advancing exaggerated claims of ten minutes and more for the duration of the tremors.

The procedure for observing seismic activity spread remarkably quickly. The first group to be affected by this comprised provincial academicians and their official and unofficial correspondents. The documents collected by the Académie des sciences demonstrate, however, that these observational procedures were also taken up by social groups much more diverse in character than the usual correspondents of academicians. At the time of the seismic activity of 1755-1756, and thereafter, numerous individuals recorded observations and presented them to scientific bodies. Considering the whole of Philippe Buache's files, we find in no particular order correspondence from a Cordelier nun from the convent of Saint-Florent-le-Vieil in Maine-et-Loire, an auxiliary engineer of the highways department, an infantry lieutenant, another engineer from Quebec, provincial academicians and their correspondents, Carthusian monks, Cordelier monks from Salins, an adolescent aged between fifteen and sixteen with only basic literacy, a physician and correspondent of the Académie des sciences, a member of the Portuguese oratory, a Jesuit, a merchant from Martinique, a mining franchisee, and a factory inspector from the Saint-Gobain glassworks.[52] Equally, among the participants in the Académie de Rouen's 1756 essay competition, there were a cleric from the Evreux chapter, a Breton country squire, an architect from Mamers, and the royal prosecutor of Azay-le-Rideau.[53]

Unjustly neglected, the physics debate over earthquakes ought to be regarded as one of the first popular debates of Enlightenment science, raging just as fiercely, if not more so, than the ballooning mania of the 1780s. Neither were the lower orders excluded from this geological craze, even in the countryside. In 1778 in Le Mans, a marvellous "mécanique de figures mouvantes qui montre la ville d'Orléans assiégée par les Anglais et délivrée par 'Jeanne d'Arque' et

[51] Académie Royale des Sciences, Registre des Procès-verbaux des séances, 4 août 1756, vol. 75, fol. 455-56.
[52] Bibliothèque Nationale de France, ms. n.a.fr. 20236 et 20237.
[53] Bibliothèque municipale de Rouen, fonds de l'Académie, MS C 20, "Concours de 1756".

le tremblement de terre de Lisbonne"[54] was set up. This contraption, the mechanics of which remain a mystery, must have travelled between the regional fairs. Already the tremors of the 1750s were accompanied by exuberant displays of collective interest up and down the kingdom. The *Journal encyclopédique* paints the following picture:

> Tous les physiciens sont occupés aujourd'hui à chercher la véritable cause des tremblements de terre; les Académies attachent un prix à cette découverte; les Ecoles ne retentissent que des causes de ce cruel phénomène; dans les cercles les plus brillants où l'on ne s'occupe ordinairement que des choses les plus frivoles, on en fait la matière de la conversation; l'ignorant même ose en parler, & suivant de loin le savant qui sait s'arrêter à propos, il se perd bientôt dans les gouffres de la terre entrecouverte de ses pas. Tout le monde en un mot veut pénétrer ce terrible secret de la nature.[55]

Yet if there were seismophiles, there were also seismophobes. A scholar from Bordeaux, one M. de Romas, found this out to his cost in 1759 when he sought to experiment publicly with electricity in a corner of the Jardin Royal. The earthquake of 10 August that year provided him with the opportunity to demonstrate that electrical theories could explain seismic motion. His audience, however, preferred to believe that his invocations were somehow linked to the catastrophe which had just struck the city, and rose up against the physicist whose apparatus only just escaped destruction.[56] Similarly, in 1756 Anne-Henriette de Briqueville attributed the increased number of tremors to the actions of electrical machinery which, she maintained, formed unstable concentrations of sulphur in the Earth's crust.[57] In the same year, one of the essays submitted to the Académie de Rouen denounced the increasingly frequent use of electrical apparatus which, it claimed, stirred up electrical matter within the Earth and connected it

---

[54] Jean Quéniart, *Culture et société urbaines dans la France de l'ouest au XVIIIe siècle* (Paris: C. Klincksieck, 1978), p. 126.

[55] "Les tremblemens de terre attribués à l'électricité", *Journal encyclopédique*, (1er mai 1756).

[56] André Grellet-Dumazeau, *La Société bordelaise sous Louis XV et le salon de madame Duplessy* (Bordeaux: 1897), pp. 262-63.

[57] Anne-Henriette de Bricqueville, *Réflexions sur les causes des tremblements de terre, avec les principes qu'on doit suivre pour dissiper les orages tant sur terre que sur mer* (Paris: 1756). Compte rendu dans le *Journal de Trévoux* (déc 1756), pp. 3012-16.

to electricity occurring in the atmosphere. Its author concluded with the wish "que les expériances de l'électricité ne fussent pas si fréquantes et même pour parler naturellement, qu'on les supprime entièrement, car pour quoy vouloir contraindre les éléments à produire des effets contraires à ce qui leur est naturel".[58] The truly Europe-wide impact of the events in Lisbon is cast in a peculiarly new light by such responses.

Conclusion

With the Lisbon disaster, natural catastrophe became an historical event. It was no longer one among many other interchangeable signifiers of Nature, or of the natural world, grounding its meaning in an ahistorical transcendence. Catastrophe here moved away from its etymological sense of a theatrical denouement which promises at once a new beginning. It became instead a break in the ordinary run of things, demarcating a "before" and an "after". The secularization of catastrophe had less to do with a wholesale rejection of religious connotations that it did with the transformation of catastrophe into a contingent event occurring at a specific time and in a specific place.

The historicization of natural catastrophe established a new relation with the present. For the first time in centuries the degree of risk deemed acceptable by mankind had changed; in escalating a sense of risk in society at large, the Lisbon earthquake of 1755 opened the way for other dangers to become quantifiable risks, from volcanic eruptions to the incidence of lead poisoning. For the first time too, the impact of the events in Lisbon brought into direct contact the localized experience of disaster and the nationwide debates about it, thereby giving a voice to sections of the population who previously had had little or none. The sense, then, of a proliferation of seismic events gave rise to a new creative tension between the heightened awareness of man's vulnerability and the renewed promise of his eventual triumph over nature. The politicization of the catastrophe in Lisbon also collapsed the safe distance that had hitherto separated sovereigns from natural disasters; these latter would henceforth represent crises capa-

[58] Bibliothèque municipale de Rouen, fonds de l'Académie, MS C 20, "Concours de 1756 sur les tremblements de terre, mémoire n° 10".

ble of shaking the legitimacy of political power to its very foundations. Ultimately, natural catastrophe also became a socal issue, giving rise to contrasting and concurrent interpretations of its meaning.

Contrary to what the philosopher, Michel Serres, wrote in the aftermath of the Asian tsunami of Christmas 2004,[59] the Enlightenment never championed a triumphalist reading of science's ability to overcome natural dangers, a reading which would be called into question by a series of subsequent disasters. The debt we owe to Lisbon lies elsewhere, specifically in the inscription of natural catastrophe in the historical process, thereby founding the western conception of historical progress as a gradual liberation of humankind from the twin dangers of Nature and Evil. 1755 marks the first symbolic victory by which natural catastrophe is transformed into a philosophical and cultural event, stripping it in the process of its most mysterious elements.

Yet – as we are discovering today – increasingly sophisticated and complex societies generate a new order of vulnerability, often greater than that experienced in the past. Concentrations of several million people, equipped with expensive technological systems and an ultramodern network of communications, are ultimately fragile things. Risk and death are part of the normal functioning of any social system, hence the urgency of being aware of them and of protecting oneself in advance. The utopian idea, often seen as accompanying the Lisbon disaster, that we might one day triumph over nature, has to be abandoned. But it would be equally wrong not to strive to do more, for the contemporaries of the Lisbon earthquake clearly showed that the response to any natural catastrophe is an ongoing project, work in progress, a goal to be aimed for, while always remaining aware of one's vulnerability. On this point, we still have a lot to learn from the events in Lisbon over 250 years ago.

Translated by David McCallam

[59] Michel Serres, "Interview", *Le Figaro*, 6 juin 2005.

Images of the Earth, Images of Man: The Mineralogical Plates of the *Encyclopédie*

Rebecca Ford

Abstract: *Despite critical attention having been paid to the illustrative plates of the massive* Encyclopédie *project, this attention has been mainly directed at the illustrations of trades, crafts and arts. The chapter examines the neglected plates on natural and mineralogical phenomena, largely drawn to accompany the texts of d'Holbach on the earth sciences. Not only does this approach enhance and complement our understanding of the* Encyclopédie *project itself, it also marks a significant stage in the birth of the earth sciences as a distinct scientific discipline in eighteenth-century France.*

"Il faut tout examiner", wrote Diderot in his *Encyclopédie* article "ENCYCLOPEDIE", "tout remuer sans exception et sans ménagement: oser voir."[1] While here discussing the *Encyclopédie*'s attitude to knowledge in general, Diderot's statement is no less applicable to the project's eleven plate volumes. Presented in the 1749 *Prospectus* as one of the most innovative elements of the *Encyclopédie*, their appearance from 1762 onwards assured the project's ongoing presence in the public mind while the final ten volumes of text were being

[1] Denis Diderot, "ENCYCLOPÉDIE", *Encyclopédie; ou, Dictionnaire raisonné des sciences, des arts et des metiers*, ed. by Denis Diderot and Jean le Rond d'Alembert, 33 vols (Paris: Le Breton, Briasson, Durand, David l'ainé, 1751-80; repr. Stuttgart: Frommann, 1966-67), V, pp. 635^{r}-48^{v} (p. 644^{v}). Further references to the *Encyclopédie* are given after quotations in the text. Spelling and punctuation are as given in the *Encyclopédie*, although ampersands have been changed to "et".

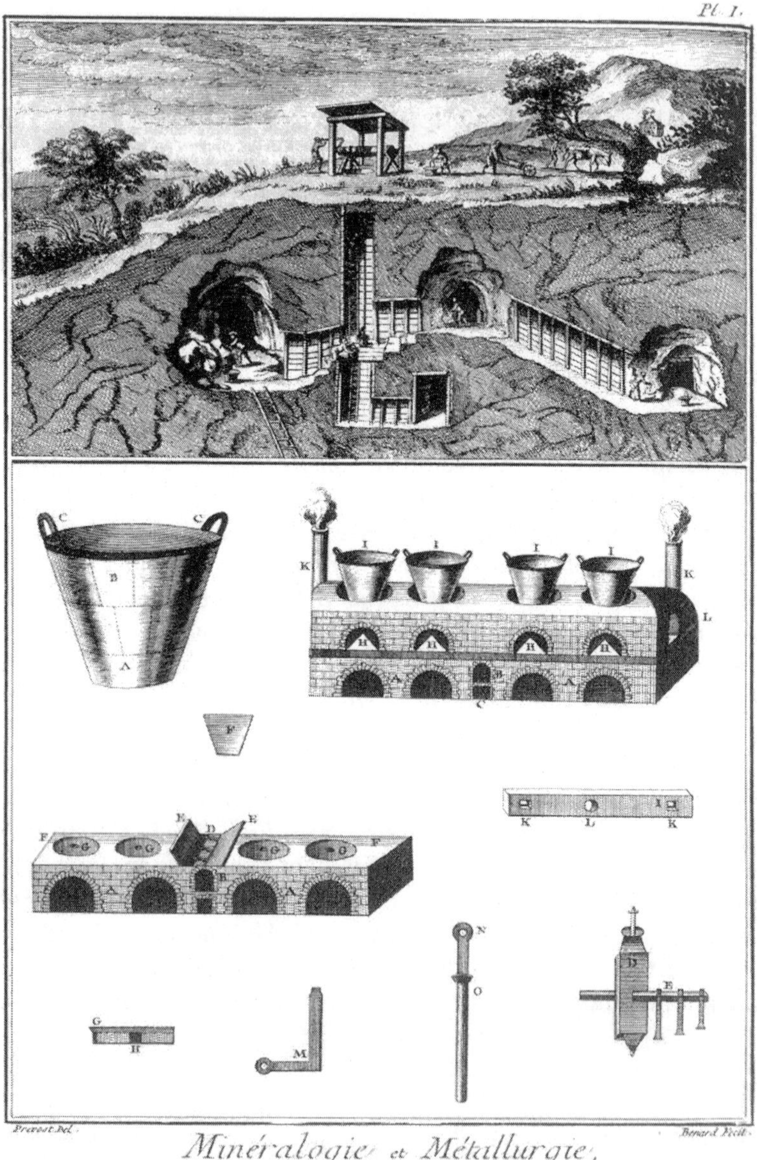
Pl. I.
Prevost Del.
Benard Fecit.
Minéralogie et Métallurgie,
Or, Coupe d'une Mine, Tirage de la Mine.

Fig. 1

clandestinely prepared for publication.[2] And although they seemed less openly subversive than the work's text volumes, the plates nevertheless contributed to the *Encyclopédie*'s aim of presenting the reader with a clear and detailed account of all branches of human knowledge. The most well-known of the *Encyclopédie* plates are those depicting the practical arts. Headed by a vignette of the workshop or shop front, and followed by exploded diagrams of machinery and equipment, successive plates display the processes, equipment and products of each art in ever-increasing detail, with the aim of enabling the reader to comprehend even the most complex of procedures. Thus in the series of plates dedicated to the refinement of gold, we find vignettes taking the reader from the extractions of the ore from the mine through to its calcinations at the forge, each accompanied by an orderly presentation of the equipment involved (figs 1&2). Each individual item, each human figure, is labelled and explained in the accompanying notes. Nothing is mysterious or haphazard: all is rational, ordered and comprehensible, laid out before the reader's inquisitive gaze.

The mineral kingdom too finds representation in the *Encyclopédie*'s pictorial account of human knowledge, namely in the sixth plate volume, published in 1768 and dedicated to natural history in general; the section on mineralogy itself was overseen by the baron Paul Thiry d'Holbach following a disagreement between Diderot and Pierre Daubenton, older brother of the naturalist Louis-Jean-Marie Daubenton, and who had originally been assigned the editorship of the natural history plate volume.[3] Anonymously introduced to the *Encyclopédie* readership as "une Personne, dont l'Allemand est la

[2] "Prospectus de l'*Encyclopédie*", in *Diderot: Œuvres Complètes*, V, *Encyclopédie, I*, ed. by John Lough and Jacques Proust (Paris: Hermann, 1976), 83-130 (pp. 101-3).

[3] Mineralogy is used here in its eighteenth-century sense, that is, the study of the Earth's composition and history, and the study of individual elements of the Earth's crust such as minerals, rocks, fossils etc. *Diderot: Correspondance*, ed. by Georges Roth and Jean Varloot, 16 vols (Paris: Editions de Minuit, 1955-70), IX (1963), pp. 29, 31, to Le Breton (4 March 1769); John Lough, *Essays on the "Encyclopédie" of Diderot and d'Alembert* (Oxford: Oxford University Press, 1968), pp. 116-17. For information about both Daubenton brothers, especially Louis-Jean-Marie Daubenton, renowned naturalist and the main contributor on natural history to the *Encyclopédie*, see Frank A. Kafker and Serena L. Kafker, *The Encyclopedists as Individuals: A Biographical Dictionary of the Authors of the "Encyclopédie"* (*Studies on Voltaire and the Eighteenth Century,* 257 (1988)), pp. 92-93.

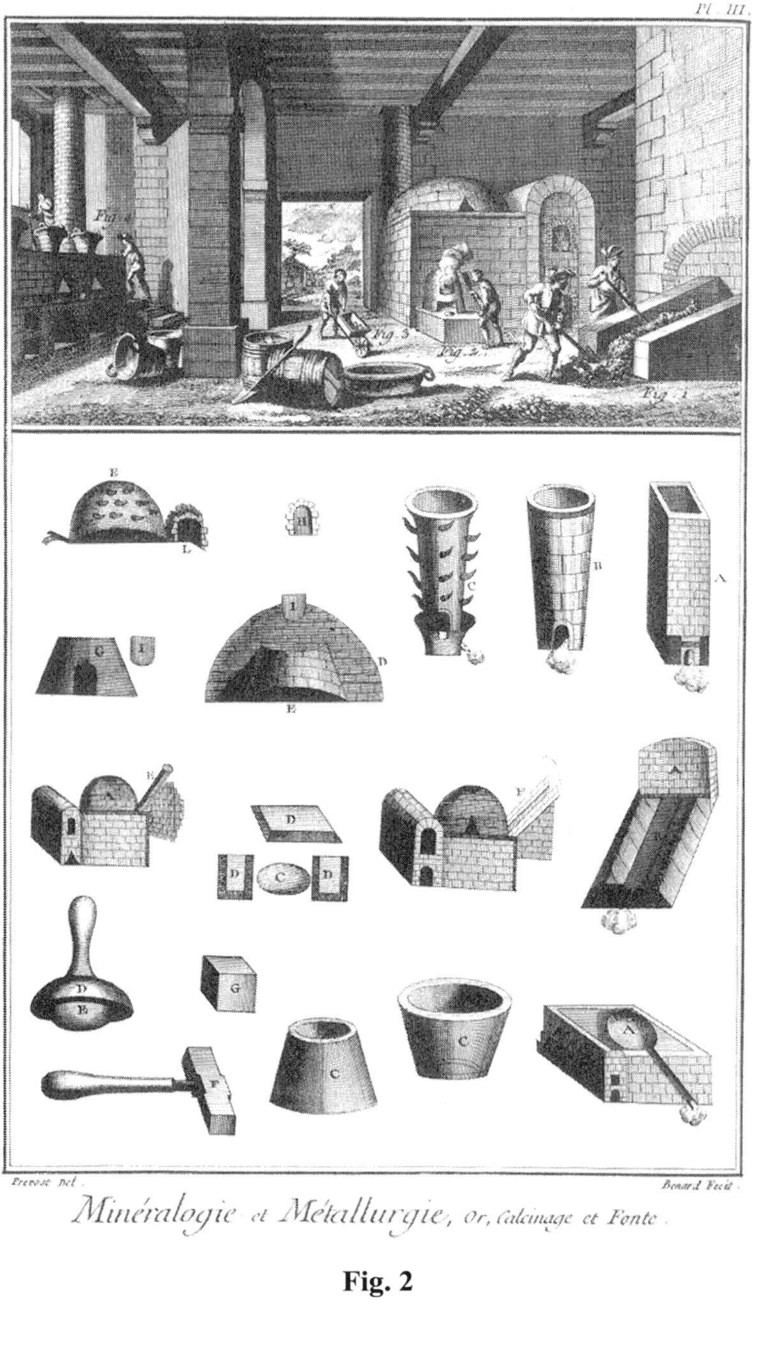

Fig. 2

Langue maternelle, & qui est très-versée dans les matières de Minéralogie, de Métallurgie, & de Physique [...] qui cultive les Sciences sans intérèt, sans ambition, & sans bruit; & qui, content du plaisir d'être utile, n'aspire pas même à la gloire si légitime de le paroître", d'Holbach was celebrated by the French reading public much less as a materialist atheist than as a disseminator of knowledge on mineralogy, supplementing his two hundred and more articles on mineralogy for the *Encyclopédie* with a series of translations of German and Swedish chemists and mineralogists.[4] As Jacques-André Naigeon wrote in his obituary of d'Holbach for the *Journal de Paris*, "C'est à lui que l'on doit en grande partie les progrès rapides que l'histoire naturelle et la chimie ont faits il y a environ trente ans parmi nous; c'est lui qui en a inspiré le goût et même la passion."[5] Foreign visitors to Paris noted the scientific flavour of discussions at d'Holbach's salon: Charles Burney commented that "I was entertained and enlightened very much by the Baron's conversation on Chymistry, Minerals, Fossils and other parts of Natural History, of which he seemed a perfect master – many

---

[4] Diderot, "Avertissement des éditeurs", *Encyclopédie* II, i. The ten scientific works translated by d'Holbach are: Johann Kunckel, Christopher Merret and Antonion Neri, *Art de la verrerie* (Paris: Durand and Pissot, 1752); Johann Gottschalk Wallerius, *Minérologie; ou, Description générale des substances du règne mineral*, 2 vols (Paris: Durand and Pissot, 1753); Johann Friedrich Henckel, *Introduction à la mineralogie; ou, Connoissance des eaux, des sucs terrestres, des sels, des terres, des pierres, des minéraux, & des métaux; avec une description abrégée des operations de métallurgie*, 2 vols (Paris: Cavelier, 1756); Christlieb Ehregott Gellert, *Chimie métallurgique, dans laquelle on trouvera la théorie & la pratique de cet art, avec des experience sur la densité des alliages des métaux, & des demi-métaux, & un abrégé de docimastique*, 2 vols (Paris: Briasson, 1758); Johann Gottlob Lehmann, *Traités de physique, d'histoire naturelle, de mineralogie & de métallurgie*, 3 vols (Paris: Hérissant, 1759); Henckel, *Pyritologie; ou, Histoire naturelle de la pyrite*, 2 vols (Paris: Hérissant, 1960); Johann Christian Orschall, *Œuvres métallurgique* (Paris: Hardy, 1760); *Recueil des memoires les plus intéressants des chymie & d'histoire naturelle, contenus dans les actes de l'Académie d'Upsal, et dans les memoires de l'Académie royale des sciences de Stockholm* (Paris: Didot le jeune, 1764); Georg Ernst Stahl, *Traité du soufre* (Paris: Didot le jeune, 1766); Wallerius, *L'Agriculture réduite à ses vrais principes* (Paris: Lacombe, 1774).

[5] Naigeon, "Lettre sur la mort de M. le Baron d'Holbach", *Journal de Paris*, 12 February 1789, quoted in Pierre Naville, *Paul Thiry d'Holbach et la philosophie scientifique au XVIIIe siècle* (Paris: Gallimard, 1943), p. 452. D'Holbach's obituary in the *Correspondance littéraire* of March 1789 similarly celebrates d'Holbach's contribution to science. *Correspondance littéraire, philosophique et critique par Grimm, Diderot, Raynal, Meister, etc.* ed. by Maurice Tourneaux, 16 vols (Paris: Garnier Frères, 1877-82; repr. Nendeln/Liechtenstein: Krauss Reprint, 1968), XV, p. 416.

of the best articles on those subjects and Metallurgy in the Encyclopédie are his",[6] although Horace Walpole, aware of the potential philosophical implications of the study of the Earth's history, was less favourably impressed:

> I sometimes go to Baron d'Olbach's, but I have left off his dinners, as there was no bearing the authors, and philosophers, and savants, of which he has a pigeon-house full. They soon turned my head with a new system of antediluvian deluges, which they have invented to prove the eternity of matter. The Baron is persuaded that Pall Mall is paved with lava or deluge stones.[7]

D'Holbach was thus the ideal candidate for overseeing the *Encyclopédie* plates on mineralogy after Diderot's dispute with Daubenton le jeune, combining as he did expertise in mineralogy with a keen involvement in the *Encyclopédie* project as a whole (as may be seen from his contribution of more radical anti-religious articles as well as from his longstanding friendship and continual support of Diderot throughout the years of the *Encyclopédie*'s publication). His well-regarded art collection, moreover, although it may well have been just one more element in the lifestyle of a wealthy *philosophe,* also suggests an aesthetic interest in the mineralogical plates above and beyond their scientific context.[8]

The section of plates on mineralogy contains plates showing individual mineral and fossil samples and numerous depictions of mining and metallurgy that conform to the presentation of the arts, as well as plates showing large-scale natural phenomena such as volcanoes, glaciers, and rock formations; plates which initially seem in

---

[6] Quoted in R. A. Leigh, "Les Amitiés françaises du Dr. Burney", *Revue de littérature comparée*, 25 (1951), 161-94 (p. 170).

[7] *The Yale Edition of Horace Walpole's Correspondence,* ed. by W. S. Lewis, Robert A. Smith, and Charles H. Bennett, 48 vols (New Haven: Yale University Press, 1937-83), xxx, *Horace Walpole's Correspondence with George Selwyn, Lord Lincoln, Sir Charles Hanbury Williams, Henry Fox, Richard Edgcumbe,* ed. by W. S. Lewis and Robert A. Smith (1961), p. 208 (to George Selwyn, Monday 2 December 1765).

[8] For details of d'Holbach's art collection, see *Catalogue de tableaux des trois écoles, Estampes en Volumes & en Feuilles, Figures de bronze, Vases de marbre, Porcelaines, Bronzes dorés, Histoire naturelle & autres objets; Formant le Cabinet de M. le Baron d'Holback ; des Académies de Pétersbourg, de Manheim & de Berlin* (Paris: Le Brun, 1789) (date of sale: 16 March 1789). A facsimile reprint of d'Holbach's art catalogue is published in *Catalogue de tableaux des trois écoles; Eléments de la morale universelle; ou, Catéchisme de la nature*, ed. by Jeroom Vercruysse (Geneva: Slatkine Reprints, 1979).

striking contrast to the better-known plates on manufacturing procedures. Instead of a vignette followed by a successive enumeration and elaboration of the tools and machinery involved, these plates consist mainly of single images, with few of the illustrative techniques of breaking down and peering used in the plates on the mechanical arts. Compared with the dynamic depiction of the processes involved in human industry, these plates may at first glance appear static, offering little understanding of the processes at work within Nature beyond an aesthetic depiction of their visible effects. Certainly, much of this has to do with the nature of the subject presented and the development of mineralogy as a science in the mid-eighteenth century. Although an *Encyclopédie* article and a plate are dedicated to the art of subterranean geometry as a means of mapping the contents of the Earth's crust, cross-sections of strata were as yet extremely rare. Although some geological cross-sections are to be found in seventeenth-century texts, these tended to illustrate cosmological or cosmogonical theories rather than observation-based representations of a particular locale; it was not until the mid-eighteenth century that stratographical depictions of particular regions began to be published in works on the study of the Earth.[9] In fact, one of the earliest examples of such images is to be found in Johann Gottlob Lehmann's *Traités de physique, d'histoire naturelle, de minéralogie et de métallurgie*, published in 1756 and translated into French by d'Holbach himself in 1759. Similarly, the notion of the Earth having a traceable history was still relatively new, and the type of time-lapse images used in the *Encyclopédie* to show various stages of production in the arts found their closest parallels with the history of the Earth again in works of cosmology, such as the frontispiece of Thomas Burnet's *Telluris Theoria Sacra* of 1689, which depicts the history of the Earth as conceived from a Christian perspective, with the Earth being shown in the various stages suggested by seventeenth-century theology and Biblical accounts, moving from its initial primeval state through a paradisiacal stage without waters or mountains, followed by the Deluge and then the Earth's current state, to the anticipated conflagration and the subsequent creation of a

[9] Charlotte Klonk, "Science, Art, and the Representation of the Natural World", in *The Cambridge History of Science, IV: Eighteenth-Century Science*, ed. by Roy Porter (Cambridge: Cambridge University Press, 2003), pp. 584-617 (pp. 612-13).

New Earth and the New Heavens.[10] Such images, and their cosmological associations, were unlikely to have found favour with the materialist d'Holbach.

Different as they are from *Encyclopédie* images of the practical arts, it might be asked then, in what ways these images can be seen to fit with the *Encyclopédie* project and its attitude to knowledge. Although they might seem to be something of an "optional extra" to the plates on the arts of which Diderot was so proud, the plates on large-scale natural phenomena are nevertheless valuable firstly in the snapshot they offer of the ways in which the eighteenth century related to the mineral kingdom, and secondly in that despite their difference from "typical" plates on the arts, such images can nevertheless be understood as being closely linked to the *Encyclopédie*'s attitude to knowledge and to d'Holbach's articles on mineralogy.

The varied provenance of these plates reveals in itself the variety of contexts in which the eighteenth century related to the natural world.[11] Natural history attracted a huge following in the mid-eighteenth century, and as Louis-Jean-Marie Daubenton, Buffon's assistant at the Jardin du Roi, noted in his own article on natural history for the *Encyclopédie*, the subject was open to study by a variety of people in a variety of ways:

> L'*Histoire naturelle* est inépuisable; elle est également propre à exercer les génies les plus élevés, et à servir de délassement et d'amusement aux gens qui sont occupés d'autres choses par devoir, et à ceux qui tâchent d'éviter l'ennui d'une vie oisive; l'*Histoire naturelle* les occupe par des recherches amusantes, faciles, intéressantes, et variées, et par des lectures aussi agréables qu'instructives. Elle donne de l'exercice au corps et à l'esprit; nous sommes environnés des productions de la nature, et nous en sommes nous-mêmes la plus belle partie. On peut s'appliquer à l'*Histoire naturelle* en tout tems, en tout lieu et à tout âge. ("Histoire Naturelle", VIII, 228)

---

[10] Stephen Jay Gould, *Time's Arrow, Time's Cycle*, 2nd ed. (London: Penguin, 1988), pp. 21-59.

[11] Although no study has undertaken so far to determine the engraver and/or original source for every image included in the *Encyclopédie*'s eleven volumes of plates, a considerable amount of research has been done by scholars such as Richard N. Schwab, Walter E. Rex, John Lough, and Madeleine Pinault into those *Encyclopédie* contributors involved in the execution of the plates. See Schwab, Rex, and Lough, *Inventory of Diderot's "Encyclopédie"*, *Studies on Voltaire and the Eighteenth Century*, 223 (1984); Pinault, "Diderot et les illustrateurs de l'*Encyclopédie*", *Revue de l'Art*, 66 (1984), pp. 17-38.

In keeping with Daubenton's survey of the variety of ways and contexts in which the eighteenth century approached nature, the plates on mineralogy are taken from a variety of sources. Some images were lifted from a growing body of literature on natural history available to the general reading public, although such borrowing sometimes gave little regard to the original context of the images used. A plate which purports to show the "Glaciers de Bermina chez les Grisons", for example, was apparently drawn from the frontispiece of Gottlieb Siegmund Gruvier's *Die Eisberge des Schweizerlandes* which in fact depicts the waterfalls at Staubbach, Lauterbrunnen.[12] If works on natural history were a significant means by which the eighteenth century engaged with nature, a more direct approach is reflected in the plates depicting volcanoes. Alongside growing scientific examination of volcanoes, a trip to Vesuvius was a key stop for Grand Tourists, and volcanoes also occupied a central place in eighteenth-century European painting, dramatic night-time explosions offering artists the opportunity to use the contrast of light and shade to spectacular effect (fig. 3).[13] Eighteenth-century visitors to Italy often bought paintings of Vesuvius and the *Encyclopédie* plates depicting the volcano's eruptions are highly reminiscent of such images. Other *Encyclopédie* images, however, reflect a more scientifically motivated interest: in 1766 the artist Jean-Jacques de Boissieu, accompanying the duc de la Rochefoucauld and the geologist Nicolas Desmarest on a tour of Italy and France, took a number of drawings of Auvergne rock formations which he later submitted to Diderot for inclusion in the *Encyclopédie* (fig. 4).[14] These images, with their precise observation of the rocks' detail, display a concern for a scientific, above and

[12] "Histoire naturelle, Règne mineral, Cinquième collection, Glaciers, Planche II", XXVII; Pinault, "Diderot et les illustrateurs", p. 25.

[13] Pinault, *The Painter as Naturalist: From Dürer to Redoute*, trans. by Philip Sturgess (Paris: Flammarion, 1991), pp. 255-59; Ian Jenkins and Kim Sloan, *Vases and Volcanoes: Sir William Hamilton and His Collection* (London: British Museum Press, 1996); Barbara Maria Stafford, *Voyage Into Substance: Art, Science, Nature, and the Illustrated Travel Account, 1760-1840* (Cambridge, MA: The MIT Press, 1984), p. 249. Stafford argues that there was a clear difference between such images motivated by "an ostentatious display of artistic effects" and the studies undertaken by amateur and professional volcanologists.

[14] Marie-Félicie Perez and Madeleine Pinault, "Three New Drawings by Jean-Jacques de Boissieu", *Master Drawings*, 23-24 (1987), 389-95 (pp. 389-90); Pinault, *The Painter as Naturalist*, p. 255.

Minéralogie, 6me Collection Volcans
Histoire Naturelle, Volcans, Eruption du Vesuve en 1754.

Fig. 3

Minéralogie, 6me Collection.
Histoire Naturelle, Base d'une Bute toute composée de Prismes articulés, sur laquelle etoit située l'ancien
Chateau de la Tour d'Auvergne, à côté de laquelle on a ajouté la Vue du Pavé naturel qui recouvre une grande place, forme où se tiennent les Foires de cette petite Ville.

Fig. 4

beyond a purely aesthetic or illustrative, appreciation of natural phenomena.

In fact, the extensive notes accompanying these plates constitute the first publication of Desmarest's theories of the volcanic origin of basalt.[15] And art and science are also tightly interwoven in the plate entitled "Pavé des Géans" (fig. 5). Originally painted by the artist Susannah Drury, in 1740 this depiction of the Giant's Causeway in Northern Ireland won the £25 premium for landscape painting for the Dublin Society before eventually being incorporated into the *Encyclopédie*. Drury's careful attention to the detail of the rock formation (even if the scale of the whole is somewhat enlarged) placed her image within the context of science as much as in that of art: Desmarest, for example, referred to Drury's image in the exposition of his theory of the volcanic origin of basalt.[16]

In the plates on mineralogy, then, Nature is represented in the context of both leisure interest and scientific endeavour, a spectacle to be admired and valued for its aesthetic value as well as an object for scientific research. The various contexts in which the eighteenth century related to nature are also, however, made evident within the images themselves by the presence of human figures; and it is these figures, and the role they play, which tie the images more closely to the *Encyclopédie* project itself.

At first sight, however, the presence of humanity in the plates is easily overlooked. Even without their titles, there is little question as to what the plates are purporting to show. Dominating each image as they do, the glaciers, volcanoes and rock formations presented by the *Encyclopédie* are clearly the main focus of their respective plates. The human figures in the foreground, by contrast, initially go unnoticed; yet despite their tiny size and seeming irrelevance to the main focus of the plates, these human figures do play a role in the *Encyclopédie* reader's apprehension of the phenomena shown.

---

[15] Perez and Pinault, p. 392. On the construction of the "scientific gaze", see Stafford, *Voyage Into Substance,* pp. 31-35.

[16] Anne Crookshank, *The Painters of Ireland, 1660-1920* (London: Barrie and Jenkins, 1978), pp. 62-27; Martyn Anglesea and John Preston, "A Philosophical Landscape: Susannah Drury and the Giant's Causeway", *Art History,* 3 (1980), 252-73; Nicolas Desmarest, "Sur l'origine & la nature du basalte à grandes colonnes polygonales, déterminées par L'histoire naturelle de cette pierre, observée en Auvergne", *Mémoires de l'Académie des Sciences*, vol. 87 (1771), pp. 705-75.

Fig. 5

Firstly, the presence of humanity in plates devoted to geological phenomena may be seen to act as the measure of Nature itself, drawing on both longstanding traditions and new aesthetic concerns in landscape art. The image of the Grindelwald glacier (fig. 6), with the tranquil human scene played out in the foreground, evokes a topographical tradition in which human settlement or activity was a central concern alongside the delineation of features of a particular locality.[17] Other images, however, move closer to the tradition of the sublime in landscape painting through the way in which the tiny human figures serve not to tame the natural scene but to emphasize its drama and majesty (fig. 3). And the dual conception of humanity's place in the universe evoked by these two uses of human figures in landscape images is mirrored in the *Encyclopédie*'s conception of knowledge, in which the overwhelming immensity of phenomena to be explored nevertheless finds its justification and end in mankind itself:

> Une considération, surtout, qu'il ne faut pas perdre de vue, c'est que si l'on bannit l'homme ou l'être pensant et contemplateur de dessus la surface de la terre; ce spectacle pathétique et sublime de la nature n'est plus qu'une scène triste et muette. L'univers se tait; le silence et la nuit s'en emparent. Tout se change en une vaste solitude où les phénomènes inobservés se passent d'une manière obscure et sourde. C'est la présence de l'homme qui rend l'existence des êtres intéressante;

[17] Klonk, "Science, Art, and Representation", p. 594.

Fig. 6

et que peut-on se proposer de mieux dans l'histoire de ces êtres, que de se soumettre à cette considération? Pourquoi n'introduirons-nous pas l'homme dans notre ouvrage, comme il est placé dans l'univers? Pourquoi n'en ferons-nous pas un centre commun? Est-il dans l'espace infini quelque point d'où nous puissions, avec plus d'avantage, faire partir les lignes immenses que nous nous proposons d'étendre à tous les autres points? ("ENCYCLOPÉDIE", V, 641^{r})

When considered in the light of their related articles, the human figures of the plates function as the visual correlative of the articles' authors; their presence palpable to differing extents in different plates and articles, both the authors and the figures in the foreground of the images serve to frame and contextualize the subject, and to direct the reader's attention to salient points of interest. Thus, in one of the plates on volcanic eruptions (fig. 3), the human figures are a central element in the circular movement of the image. Their pointing hands direct the viewer's gaze to the centre of the eruption, and the lava flow which emanated from it curves down and around to the human figure, which begin the circle again. Although dwarfed by the volcanoes' eruptions, and seemingly extraneous to the drama of the natural scene, the recurrent impact of the image is dependent upon the tiny human figures. It is the presence of mankind that allows the image to "speak": even if the rationalizing action of the human mind is not as visibly at work as in the breaking-down of machines and tools in the plates on the mechanical arts, it is the human presence in the plates on mineralogy which help to lend the phenomena their significance, and justify their very inclusion in the *Encyclopédie*.

That this contextualizing of natural phenomena within the framework of human activity was sometimes deliberate, and not simply the result of borrowing ready-drawn images from other sources, is evident from one of the plates depicting Auvergne basalt (fig. 4). De Boissieu's original drawing was of the rock formation itself, devoid of any human element; the *Encyclopédie* plate, however, places this scientific observation firmly within a human context. Beside the two men in the bottom right-hand corner of the image, themselves standing in front of a cottage butting onto the rock face, a naturally-formed courtyard has been added to the foreground, and the plate's subtitle informs the reader not only that this courtyard is the location for the local village's annual fairs, but also that the three rocks atop the basalt columns are the remains of "l'ancien Chateau de la Tour d'Auvergne". In this way, Nature is humanized; and while the ruins of the old castle serve as a reminder of humanity's ephemeral nature when compared to the rocks' resilience to the ravages of time, man's ability to stamp his presence on the natural world and to adapt it to his own ends is made abundantly clear.

This appropriation of nature by man is most clearly visible, however, in the plate showing the Giant's Causeway (fig. 5). Although what immediately seizes the reader's attention is the rock face, the image is soon seen to teem with human life, the various groupings of people representing the different ways in which humans colonized this vast natural edifice. In the left foreground two men examine individual rocks; on the right-hand side of the image is a small group of people gathering food or driftwood from the rock pools; and in addition to these two groupings engaging with Nature as a source of knowledge of physical sustenance, there are a number of other figures who seem to envisage the Giant's Causeway as a consumer product, a monument to be visited and admired, but little more. On the left of the image stands a man contemplating the vast rock face: although dwarfed by the basalt columns, his field of vision neatly encompasses the entirety of the facade, and his position is matched by the pair of figures on the other side of the edifice. In this way, the awesome basalt columns which initially seem to dominate the image are appropriated and cut down to a manageable size by human vision; and completing this process, a group of people converse at the very foot of the rock face, seemingly undaunted by its size and treating it almost as little more than a convenient backdrop to the important business of social interaction. This framing of the natural spectacle within a primarily social, rather than scientific context may seem to be at odds with a rational, scientific approach to Nature promoted elsewhere in the *Encyclopédie* – d'Holbach's articles on mineralogy, for example, tend to de-emphasize the aesthetic and social contexts of rock and fossil specimens and large-scale natural phenomena in favour of their composition, causes and uses – yet somewhat paradoxically, it is the seemingly superficial appreciation of nature's spectacle depicted here which may itself provide a link between the plates on geological phenomena to those on the practical arts. Roland Barthes's analysis of the *Encyclopédie* plates argues that the vignettes opening the plates on the arts and trades show the moment of completion and consumption of the product. Such a reading may thus allow us to tie in the plates on mineralogy with their more "typical" companions. Following Barthes, this plate reveals the moment of the "completion" of the natural spectacle through its "consumption" by man.[18] The natural spectacle, while undiminished in size or impressiveness, is nevertheless open to exploitation by man, and reveals the eighteenth century's faith in man's ability to make sense of and master the natural world and his place within it to his own ends.

Such mastery of nature however, as may well be imagined, comes for the *Encyclopédie* primarily from a scientific understanding of the phenomena involved; and it is such a scientific understanding of the natural world, rather than an aesthetic appreciation of nature's spectacle, which forms the closest link between text and image in the *Encyclopédie*. This may be seen again most clearly from the plate on the Giant's Causeway and its related article, d'Holbach's "Pavé des Géans". D'Holbach's article is for the majority a precise and scientific description of the geometrical conformation of the basalt columns and

[18] Roland Barthes, "Image, raison, déraison", in *Le degré zéro de l'écriture; suivi de Nouveaux Essais critiques* (Paris: Editions du Seuil, 1972), pp. 89-105 (p. 92).

their constituent parts, referring specifically to this plate and to the individual rocks shown in the foreground of the image:

> Chaque morceau ou jointure a dans son milieu une partie convexe ou une éminence qui s'adapte parfaitement à une partie concave d'une autre articulation, et ainsi de suite: de cette manière chaque articulation a une convexité d'un côté, et une concavité de l'autre; cette convexité et cette concavité sont garnies d'un rebord qui a autant d'angles que la colonne a de côtés, et qui s'engrainent exactement sur la concavité et sur les angles de l'articulation suivante. On peut voir dans la *Planche, fig. A,* que ces articulations forment comme une couronne antique. (XII, 195)[19]

The plate thus lends visual meaning to the precise geometrical language of the article, and indeed, one of the changes made to Drury's original image in its adaptation for the *Encyclopédie*, was the redrawing of the articulations of these rocks with greater clarity, better depicting the way in which they fit together to form one basalt column.[20] But the rocks are not just presented on their own: they are examined within the image itself by two figures, and their own examination of the rocks may therefore be seen as an emphatic representation of the scientific examination of Nature voiced by d'Holbach in his article; although many approaches to Nature are visible in the Pavé des Géants plate, it is the scientific approach which is privileged through its role as bridge between text and image. And moreover, the Giant's Causeway is more than just a backdrop for the scientific examination of nature; it is itself a demonstration of d'Holbach's vision of how science should be undertaken. Remarkably dismissive of the cabinet as a route to knowledge, for d'Holbach it is only through a firsthand, physical engagement with Nature that true knowledge of the mineral kingdom may be found:

> Les spéculations tranquilles du cabinet, les connoissances acquises dans les livres ne peuvent point former un *minéralogiste*; c'est dans le grand livre de la nature qu'il doit lire: c'est en descendant dans le profondeurs de la terre pour épier ses travaux mystérieux; c'est en gravissant contre le sommet des montagnes escarpées; c'est en parcourant différentes contrées, qu'il parviendra à arracher à la nature quelques-uns de ses secrets qu'elle dérobe à nos regards. ("MINÉRALOGIE", X, 542)

[19] Attributed to d'Holbach by Schwab, Rex and Lough, *Inventory*, 91 (1972), p. 758.
[20] Crookshank, *The Painters of Ireland,* p. 67.

These plates on mineralogy, then, can be seen to represent the variety of ways in which the eighteenth century related to the natural world. The relationship between man and nature operated within the contexts of industry, leisure activity, and scientific endeavour, and thus Nature was variously a resource to be managed, a spectacle to be admired, and an object to be examined. Common to almost all of them, however, is the awareness that although the natural world is indeed awesome it is nevertheless entirely open to mankind's investigation and mastery. It is thus the *Encyclopédie*'s use of landscape and its inclusion of humanity, rather than the ostensibly rationalistic breaking-down of machines and processes used in depictions of the mechanical arts, that offers the best visual representation of this physical, direct experience of nature and, by extension, the recognition of mankind's need to engage directly with not only Nature but the sources of knowledge itself.

Peat Bogs, Marshes and Fen as Disputed Landscapes in Late Eighteenth-Century France and England

Ian D. Rotherham and David McCallam

Abstract: *Buffon and the* Encyclopédistes *had a certain impact on the contemporary appreciation of peat bogs and fenland as a particular form of landscape; but, as this chapter demonstrates, the daily management of, and interaction with, such bog landscapes benefited much more from the migration of practices and ideas between countries and regions of Western Europe. Clear examples of this given here are the reception of Arthur Young's work in France and La Rochefoucauld's study of English Fenland draining and irrigation techniques in the 1780s.*

In early modern Europe marshes and fens were commonly perceived as places of fear and loathing. Shrouded in King Lear's "fen-sucked fogs",[1] they were unsafe and evil wastes where drowning was common, where one could be poisoned by the corrupted waters or the foul air, contracting the "ague" or marsh malaria which until the late nineteenth century was literally thought to originate in the bad air (malaria) of these stagnant wetlands. Other dangers included getting lost in their featureless expanses, or flooding destroying nearby cultivations and manufactures. In 1629 one tract urging the draining of the East Anglian Fens described them as sunk in "water putred and muddy, yea

[1] William Shakespeare, *King Lear*, II, iv, 169 (New York: Bantam Books, 1964), p. 82.

full of loathsome vermin; the earth, spuing, unfast and boggie".[2] In France, swathes of swamp in the Camargue, the Dombes, the Landes and Sologne remained undrained and infected, with mortality rates in the last of these areas still standing at fifteen percent above the national average at the end of the eighteenth century.[3] These natural dangers were compounded in places where cities such as London polluted the surrounding marshlands or built slums on them. Contemporary observers of bogs and fen often imputed the abhorrent nature of the physical environment to the moral character of its inhabitants. Hence Arthur Young, writing in 1799, explains the depredations of sheep thieves in Lincolnshire by stating that "so wild a country nurses up a race of people as wild as the fen".[4] Certainly, in the second half of the eighteenth century, Romney Marsh swarmed with armed smugglers, as did parts of the Landes near Bordeaux. Politically too, wetlands provided a refuge for outlaws and a site of violent resistance to government, a breeding ground for both the anti-drainage violence of the seventeenth-century "fen tigers" and the more considerable anti-revolutionary violence of the Chouans in the Vendée marshes.

However, in England, and yet more so in France, the main source of social and political conflict in the wetlands of the seventeenth and eighteenth centuries was land rights. This habitually turned on peasants challenging seigneurial attempts to appropriate, clear or drain areas traditionally considered to be common land. This wasteland was used to graze small numbers of livestock; it was also a source of water, fuel, often in the form of peat, or bedding in the form of rushes or reeds. Thus, the commons, as such lands were called, were jealously guarded by peasant communities, not least because appropriation and enclosure of commons by the local seigneur was invariably followed by taxation, by the imposition of the "cens" (quit rent) or "lods et ventes" (sales tax) levied on all who worked, tended or exploited the newly enclosed plots.[5] Peasant resistance was

[2] William Lambarde cited in Jeremy Purseglove, *Taming the Flood: A History and Natural History of Rivers and Wetlands* (Oxford: Oxford University Press, 1988), p. 25.

[3] See John McManners, *Death and the Enlightenment: Changing Attitudes to Death in Eighteenth-Century France* (Oxford: Oxford University Press, 1986), p. 10.

[4] Arthur Young, *General View of The Agriculture of the County of Lincoln* (London: 1799), p. 223.

[5] See Pierre Goubert, *La Vie quotidienne des paysans français au XVIIe siècle* (Paris: Hachette, 1982), pp. 42-43.

grounded in the belief that all uncultivated land, including waterways and marshes, was theirs by natural right, that it was to be collectively and freely enjoyed; as such, the commons often provided a major physical and symbolic bond for what Rétif de la Bretonne describes as the communal peasant "famille".[6] In France resistance usually took the form of civil or legal disobedience or the setting up of "métayages", sharecropping partnerships between landowner and land worker(s) entailing an equal division of any profits made from working the earth. Yet resistance could also be fierce and violent, as in the smashing of sluices in seventeenth-century Lincolnshire, or the reported stoning and burning in effigy of a local drainage agent on the Somerset Levels in 1769.[7] Thus, where resistance was successful, or where there had been no attempts at forced clearance and drainage, wetlands were settled and exploited largely as they had been since the Middle Ages. They were used for the summer grazing of livestock, especially cattle and sheep; they were home to geese reared for quill and duvet manufacturing; they were hunted for wildfowl, eel and fish; if surrounding cultivation needed draining, it was drained by ridge-and-furrow ploughing. And where the geology allowed, they were, of course, exploited for peat.

Peat is essentially the accumulated remains of dead plants, trees and other vegetal matter. Modern scientific analysis has found that it is ninety-nine percent organic in composition, of which at least eighty per cent is combustible when dried out.[8] In the eighteenth century, a number of theories were advanced to explain both its composition and formation. Taking as his example the peat fields of Flanders, the great naturalist Georges Louis Leclerc de Buffon concluded that peat was produced by violent sea flooding of the coastal regions which uprooted all trees and vegetation in the area. This vegetal matter was then overlain by sea waters under which it compressed and rotted. As the sea eventually receded from this part of western Europe, the peat deposits were dragged together to accumulate and concentrate

[6] Nicolas Edme Rétif de la Bretonne, *La Vie de mon père* (Paris: 1788), II, p. 82.
[7] Purseglove, *Taming the Flood*, p. 52.
[8] For a comprehensive chemical and biological analysis of peat, see R. S. Clymo, "Peat", in *Mires: Swamp, Bog, Fen and Moor: General Studies*, ed. A. J. P. Gore (Amsterdam; Oxford; New York: Elsevier Scientific Publishing Company, 1983), pp. 159-224 (p. 159).

at the lowest geographical point, that is, in the low plains of Flanders.[9] For Buffon, the key aspect of this process of peat formation was the rotting of organic matter underwater. If the same materials rotted on land, he claims, they would only produce "du terreau et du limon".[10] However, the water seals in the peat's combustible elements, that is, its share of "phlogiston", the flammable substance or essence of fire believed by contemporary science to be contained in all combustible materials and released in all acts of combustion.[11] This, maintains Buffon, would explain why peat makes such an excellent fuel once it has rotted down and been dried out. According to Buffon's theory, peat formed under sea water is better than that formed under fresh water because the bitumen and salt content of seawater increases its combustibility. Hence his confident assertion that Dutch sea-peat is best, since "elles sont pénétrées du bitume dont les eaux de la mer sont chargées".[12]

In fact, it appears to have been something of an eighteenth-century French commonplace to associate Holland with the most advanced practices of peat-cutting and peat-burning. To take but one example, Joseph de La Porte, abbé de Fontenai and Louis Domairon's astonishingly encyclopaedic *Le Voyageur françois* (1765-1795) observes that nature has provided Holland with "une terre qui, coupée en morceaux, & exposée au soleil, se durcit & brûle dans les foyers; c'est ce qu'on appelle de la tourbe".[13] He goes on to say that this Friesland peat not only serves as domestic fuel but is also used in local industries. Moreover, in a country devoid of trees and lacking coal, this "substance inflammable" is a great resource for the working poor

---

9 Georges Louis Leclerc, comte de Buffon, *Histoire naturelle, générale et particulière*, 21 vols (Paris: Imprimerie royale, 1749-1789); *Supplément. Tome cinquième* (1778), p. 470.

10 Buffon, *Histoire naturelle*, I (1749), p. 398.

11 For a fuller account of phlogiston theory and its demise, see *The Overthrow of Phlogiston Theory: The Chemical Revolution of 1775-1789*, ed. James Bryan Conant (Cambridge, MA.: Harvard University Press, 1950).

12 Buffon, *Histoire Naturelle: Supplément. Tome cinquième*, p. 472. This passage is supplementary to Buffon's original discussion of peat in Volume I of the *Histoire naturelle* (pp. 438-39) in which the author's main concern is asserting that peat is in fact a loose form of brown coal, in refutation of Claude-Léopold Genneté's claim that coal is a distinct rock type similar in its composition to clay.

13 Joseph de La Porte, abbé de Fontenai, Louis Domairon, *Le Voyageur françois, ou La Connoissance de l'ancien et du nouveau monde*, 42 vols (Paris: Vincent; Moutard; Cellot, 1765-1795), XX, pp. 323-24.

and the idle rich alike. These last burn it in their hearths, shaping its sods into little peat "châteaux", from which surprisingly multicoloured flames "offrent à la vue un spectacle amusant".[14] Yet whether it is dug and cut in Holland, England, Russia or in France, peat is more generally regarded in the eighteenth century as a fuel of the poor; it was for poor households, for instance, that it was regularly imported and sold at the Port de la Grève in Paris.[15] This association between peat and the poor is perhaps reinforced in French by the fact that the word for peat, "la tourbe", is also used figuratively to signify a "[m]ultitude confuse de peuple".[16] There is a sense here that "peuple" refers pejoratively to a congregation of the vulgar, as in Jean-Baptiste Delisle de Sales's notion of "la tourbe des voyageurs",[17] or more specifically to the lowest orders of the population, as in Louis-Sébastien Mercier's anti-clerical play *La Destruction de la Ligue* in which a fanatical Catholic remarks of the Parisian populace: "quant à cette tourbe insensible sur laquelle il y a peu de prise, faisons-lui sentir le fouet de la terreur".[18]

Nonetheless, peat fields were not immune to the Enlightenment drive to drain marshlands and bogs. Unfortunately, if these reclaimed lands were not then immediately cultivated, drainage could prove counter-productive and costly. For the exposed and desiccated peat is prone to shrinkage, wastage and oxidation occasioned by contact with the air, a process which is also accelerated by bacterial action.[19] Thus the exposed earth wastes away and erodes back into the waters from which it has been reclaimed. In eighteenth-century England, this lesson was quickly learnt, and drained peatlands were quickly ploughed, limed, manured and harrowed with a fodder crop, usually oats or potatoes. In fact, the *Edinburgh Advertiser* of May 1800 carries a detailed account of how best to cultivate "moss and peatlands" based on the experience of a certain Mr Smith of Swin-

---

[14] La Porte et al, *Le Voyageur françois*, XX, p. 324.

[15] See M. Bagot, "Sur le Charbon de Tourbe exposé en vente par la Compagnie Callias, au Port de la Grève, à Paris", *Annales de l'Agriculture françoise*, XXV (1806), pp. 46-54.

[16] *Dictionnaire de l'Académie française*, 4th ed., 2 vols (Paris: 1762), II, p. 854.

[17] J-B-C. Delisle de Sales, *De la philosophie de la nature*, 3 vols (Paris: 1770), I, p. 129.

[18] Louis-Sébastien Mercier, *La Destruction de la Ligue, ou La réduction de Paris* (Amsterdam: s.n., 1782), p. 29.

[19] Purseglove, *Taming the Flood*, pp. 12, 57-58.

ridgemuir, Ayrshire, winner of a Highland Society Gold Medal for "his extensive improvement of a large tract of Moss on his property".[20] In brief, the procedure involves cutting out master drains, eight feet wide by four feet deep, tapering to two and a half feet deep at their extremity; marking out ridges in the top soil and cutting them in one-foot-wide strips; turning them with "a gentle declivity" towards furrows, before dressing them with lime, and later dung. The dung is carried onto the moorlands by a single-horse cart along a drained and gravelled track which is also used for removing crops later from the reclaimed land.[21] Oats are suggested as a potential first crop; but in practice the Ayrshire farmer prefers potatoes.[22] This exemplary text follows the two key strictures of the "improvers" of marshes and fens: it reclaims "waste and barren grounds" and cultivates them with innovative crop rotation doing away with the need for land to lie fallow. As it suggests, the fundamental reason for this reclamation and improvement of the peat moors is to increase the "subsistence of the People", a "necessity" which the author claims is "universally felt" by his contemporaries.[23] It was certainly felt by the famous English agronomist, Arthur Young, who, writing thirty years earlier, deplored the uncultivated state of the lands bordering the canal cut through Trafford Moss. He writes: "It is a great pity that the noble advantage of a water-carriage through the heart of this moor to so fine a market as Manchester, does not induce the owners to cultivate this waste track, which might beyond all doubt be applied to numerous uses, far more profitable than yielding peat in a country so abounding in coals".[24]

We will consider later the major practical considerations driving wetland reclamation in late eighteenth-century France and England, but first it is important to explore the Enlightenment principles underpinning this process of land reform. The formation of the larger marshlands in Western Europe was generally attributed to the action of the sea. Buffon claims that a prehistoric isthmus or land-bridge

---

[20] "Agricola", "An account of the mode and expence [*sic*] of cultivating moss and peatlands", *Edinburgh Advertiser*, 2 May 1800, pp. 2-9 (p. 2).
[21] "Agricola", "An account of the mode...", pp. 4-7.
[22] "Agricola", "An account of the mode...", p. 8.
[23] "Agricola", "An account of the mode...", p. 2.
[24] Arthur Young, *A Six Months tour through the North of England*, 2nd ed., 4 vols (London: 1770-71), III, pp. 218-19.

linking the British Isles to the continent was gradually eroded by the seas on either side of it.[25] What remained of it provided a rampart against which the sea waters deposited huge amounts of sediment and vegetal debris. As the sea receded in this part of the world, these coastal deposits were both spread over large areas and rose out of the subsiding waters. As they did so, they formed the Zealand wetlands in Holland, the Romney marshes and Norfolk, Lincoln and Ely fens in England while a similar process created the Crau de Provence and Rhône delta marshes in France. These wetlands were defined by Buffon by their inability to drain water and their propensity to flood: "Lorsque les eaux qui sont à la surface de la terre ne peuvent trouver d'écoulement, elles forment des marais & des marécages".[26] That is, marshes have no *natural* means of drainage. Hence their reclamation is to be numbered among those acts which constitute man's dominion over the natural world and a further mark of his progress towards a "perfected" civilized state. Buffon's famous *Époques de la nature* of 1778 make this clear. Writing against the Rousseauist myth of a lost Golden Age of primitive humanity as well as parodying the seven days of biblical Creation, Buffon devotes the seventh and final "Époque" of his geological history of the Earth to man's ultimate triumph over his natural environment. This is his victorious humanity: "Par son intelligence, les animaux ont été apprivoisés, subjugués, domptés, réduits à lui obéir à jamais; par ses travaux, les marais ont été desséchés, les fleuves contenus, leurs cataractes effacées [...]".[27] Moreover, according to the powerful metanarrative of a unilateral global cooling that runs through Buffon's *Époques*, the practices of both wetland draining and deforestation were not merely the effects of local land economies; they were intelligent, essential acts for retarding, if not checking, the irreversible loss of the Earth's heat. Ironically for our age of global warming, Buffon maintained that marsh draining, deforestation as well as urbanization were welcome means of raising the earth's surface temperature by drying, clearing and peopling it: "c'est lui rendre de la chaleur pour plusieurs milliers d'années".[28] The localized temperature increases of Western Europe towards the end of

[25] Buffon, *Histoire naturelle*, I, pp. 586-87.
[26] Buffon, *Histoire naturelle*, I, p. 575.
[27] Buffon, "Époques de la nature", in *Histoire naturelle: Supplément. Tome cinquième*, p. 236.
[28] Buffon, "Époques de la nature", p. 240.

the eighteenth century were thus explained by the French, English and Dutch having "abattu les forêts, desséché les marais, contenu les torrens, dirigé les fleuves et défriché toutes les terres couvertes ou surchargées des débris même de leurs productions".[29]

Insofar as his narrative of the civilizing process places such a clear emphasis on agricultural advances, Buffon's enlightened earth science would seem to agree with that of the Physiocrat school for which agriculture was the basis of all "culture", be it artisanal, industrial or luxury. As Jeremy Black remarks, the Physiocrats had a point inasmuch as all contemporary manufacturing processes used only natural products and hence, at some point in the chain of production, relied on agricultural activity for their raw materials.[30] Yet Buffon's earth science was also informed by other innovative Enlightenment fields of research, namely, statistics and demography. The great naturalist's urging to drain marshes would be determined as much, then, by issues of public welfare as by geological theorizing. For pioneering demographers of the time such as "Moheau" (the pseudonym of the baron de Montyon) calculated life expectancy on the marshy plain of La Napoule to be a startling average of eight years, all age groups taken together, compared to the relative longevity of thirty-two years found in the mountainous environs of Apt.[31] As John McManners states, the succinct conclusion to be drawn from late eighteenth-century French demography was "live on mountain tops and avoid marshes".[32]

Buffon's theories of wetland formation and wetland reclamation also drew on, and fed back into, more diffuse intellectual currents in the French Enlightenment. For instance, the peasant's stubborn defence of commons, including marshland, was attacked by the same sweeping movement of thought that denounced abusive seigneurial privileges: were both not vestiges of the same depised "feudal" order that resisted all types of reform in France? Were not peasant fallows, furlong farming and ridge-and-furrow draining on a par with Gothic

---

[29] Buffon, "Époques de la nature", p. 241.
[30] Jeremy Black, *Eighteenth-Century Europe*, 2nd ed. (London: Macmillan, 1999), p. 27.
[31] McManners, *Death and the Enlightenment*, p. 100.
[32] McManners, *Death and the Enlightenment*, p. 103.

architecture – bizarre, disorderly, unenlightened practices?[33] Instead, suggested the adepts of Voltaire and Montesquieu, look to England for inspiration in order to reform both agricultural and social regimes. In terms of agriculture at least, nothing epitomizes more clearly the prevailing mood of Anglomania than François and Alexandre de La Rochefoucauld's tours of southern England in 1784 and 1785. Their father, the duc de Liancourt, Master of the King's Wardrobe, was already a keen Anglophile as well as a social reformer, having established France's first "École des Arts et des Métiers" on his estate to the north of Paris.[34] His sons followed his practical example, taking a house near Bury St Edmunds in Suffolk in January 1784 and frequenting the great English agronomist, Arthur Young, whose farm stood nearby. Their short agricultural tours of Suffolk and Norfolk in the July and September of that year, in the company of Young, convinced the brothers of the superiority of English agricultural methods: its innovative crop rotation system which did not require land to lie fallow, its comprehensive use of enclosure, its novel use of carrots and turnips as fodder crops on reclaimed fenland, and its consequent concentration of livestock farming. In fact, François de La Rochefoucauld, in his journal of this 1784 tour, explicitly links English agricultural progress to its advanced form of political government, specifically to a lack of state interference in farming, an equitable tax system which "makes itself severely felt by the rich and very little by the poor", citing also the great respect shown towards farming because "the highest in the land engage in it".[35] This last point, he notes, also has the advantage of an economy of scale: "Experiments are made on a big scale by the [noble] amateurs and they are promptly taken up by the farmers".[36] Sons of one of France's oldest noble families, it is little surprise that the La Rochefoucaulds favour this "top-down" model of agricultural Enlightenment. Yet, as François remarks, English agricultural reform works on all social levels because it is ultimately profitable for

[33] For more on this anti-"feudal" movement, see J. Q. C. Mackrell, *The Attack on "Feudalism" in Eighteenth-Century France* (London: Routledge & Kegan Paul, 1973).

[34] Norman Scarfe, *Innocent Espionage: The La Rochefoucauld Brothers' Tour of England in 1785* (Woodbridge: Boydell Press, 1995), p. 1.

[35] François de La Rochefoucauld, *A Frenchman in England in 1784: Being the "Mélanges sur l'Angleterre" of François de La Rochefoucauld*, ed. Jean Marchand, trans. S. C. Roberts (London: Caliban Books, 1995), pp. 197-98.

[36] La Rochefoucauld, *A Frenchman in England*, p. 197.

small and large landowners alike: "the more a farm is improved, the better is the return secured upon the money expended".[37]

Yet it would be wrong to presume that the profitability of wetland reclamation was an eighteenth-century realization. Already at the start of the seventeenth century, Henri IV of France had invited and paid Dutch engineers to oversee the reclamation of large areas of marsh in the Somme estuary, the lower reaches of the Seine and in Normandy. Their expertise consisted in "assécher, dessaler, drainer, construire des polders [reclaimed lowlands] avec leurs canaux".[38] The most famous of these drainage experts was Peter Bradley, the unlikely-sounding name of a Dutchman from Bergen op Zoom, who became Henri IV's Master of the Dikes, and supervised the reclamation of the great Poitevin marshes to the north of La Rochelle. Another controversial Dutchman, Cornelius Vermuyden, performed the same function in England, draining Hatfield Close, south of the Humber estuary, from 1626 to 1629. He also advanced projects for similar massive draining of the Somerset Levels, although these were never executed.[39] As Marc Bloch has suggested, these Dutch engineers did not undertake this work purely for the public good; their operations in the Fens and French coastal marshes were proto-capitalist exercises "directed by an association of technical experts and business men [...] financed by a few large business-houses, mostly Dutch".[40] Their objective was less public land reclamation than private wealth generation. As such, these wetland initiatives became the financial model for eighteenth-century reclamation enterprises in Brittany and Guienne where "companies were founded for the express purpose of financing – or indeed speculating in – land reclamation, which now also received government patronage".[41] Obviously, as we noted earlier, there was often deep suspicion, indeed open hostility, among local land workers towards these speculators and investors, these "Adventurers" as they were sneeringly called by the seventeenth-century Fenmen who smashed their sluices and pulled down their dikes.[42]

---

[37] La Rochefoucauld, *A Frenchman in England*, p. 237.
[38] Goubert, *La Vie quotidienne des paysans français*, p. 14.
[39] Purseglove, *Taming the Flood*, pp. 46-49.
[40] Marc Bloch, *French Rural History: An Essay on its Basic Characteristics*, trans. Janet Sondheimer (London: Routledge & Kegan Paul, 1978), p. 19.
[41] Bloch, *French Rural History*, pp. 19-20.
[42] Purseglove, *Taming the Flood*, p. 55.

Nonetheless, financial greed was not in itself a great enough incentive to drive the early modern processes of wetland draining, clearance and cultivation. Exploiting the general reputation of marshes and bogs as places of evil and infamy, religion and public morality were also invoked as reasons for their reclamation. Indeed, the very notion of "reclamation" – as with the later eighteenth-century idea of "improvement" – struck a strong moral chord with contemporaries, especially among the first generation of engineers and drainage experts in England, many of whom were Protestants and Huguenots fleeing religious persecution and civil strife on the Continent. Certainly, among such avid Bible-readers, the constant threat of flooding posed by the marshes and fens to surrounding agriculture and dwellings might have recalled Noah's flood, and was therefore to be checked not just on social and financial grounds, but also on religious ones too. The effects of regular flooding elsewhere in seventeenth and eighteenth-century Europe must only have reinforced this impression: Florence was heavily flooded in 1740 with the loss of many lives; Avignon was swamped in 1763 when the Rhône burst its banks; and in the winter of 1787 heavy rains and flooding swept away seed grain in Saxony, causing famine the following year.[43] In the Breton forests too, a lack of drainage expertise meant that heavy rains waterlogged and weakened the roots of trees, drowned all seedlings, and left the forest vulnerable to storm damage, a threat that was realized with devastating effect at least four times in the eighteenth century.[44]

Effective draining, then, was crucial. By 1710 the English reclaimers of the Fens had imported from Holland the technique of using windmill-powered pumps to drain the wetlands. It was not, however, until much later in the century that a further discovery and subsequent innovation in draining expanded the practice of wetland reclamation, first in Britain and then on the Continent. This discovery took place near Leamington Spa in 1764 when a local farmer, Joseph Elkington, solved the problem of underdrainage, that is, of clearing not just the surface waters from an area of wetland but of siphoning off low-level underground water tables by tapping and diverting their springs. In 1795 the British Parliament awarded Elkington the hand-

[43] Black, *Eighteenth-Century Europe*, p. 19.

[44] Andrée Corvol, "Tempêtes sur la forêt française XVIIe-XIXe siècles", in *L'Événement climatique et ses représentations (XVIIe-XIXe siècle)*, ed. E. Le Roy Ladurie, J. Berchtold, J.-P. Sermain (Paris: Desjonquères, 2007), pp. 43-59 (p. 48).

some sum of £1000 in recognition of his innovation, and asked him to survey the general application of his method of underdrainage in other parts of the country.[45] Drainage, underdrainage, and irrigation too, were then important advances in land management in eighteenth-century Western Europe. Generally, they were most effective where there were large local labour forces. This is understandable not only because drainage was a labour-intensive process, but also because it was in heavily populated areas that the demand was greatest for increased agricultural production and hence for more land on which to provide it. The population increase in Britain and France, especially after 1740, was at once the cause of wetland reclamation and a means to effect it.

In France the population rose from around 21 million in 1700 to 28.1 million in 1790.[46] This demographic pressure was reflected in governmental attitudes: the "intendants" of French regions such as Brittany, who had been close to the peasantry in the seventeenth century and supported their fight to retain common lands, were by the middle of the eighteenth century persuaded to sacrifice commons to enclosure and increased food production.[47] Increased food production was also the major incentive for fenland drainage in England in the eighteenth century. Arthur Young, writing in 1772, suggested "breaking up uncultivated lands" and "draining fens" as the chief means not only of enhancing productivity of these holdings but also of raising income for those who farmed them.[48] For the logic was straightforward: reclaimed lands could be enclosed and incorporated into the new crop rotation cycles which avoided leaving fields fallow; they could specifically be used for cultivating fodder crops which in turn would feed greater numbers of livestock which would, in their turn, produce more manure, further reducing the need for fallow and fertilizing yet greater yields on the reclaimed lands. If this was the English model, it also began to be widely adopted in France in the late eighteenth century. In Normandy, for instance, a better supply of fodder crops from reclaimed lands, including marshes, encouraged animal husbandry, especially with regard to cattle; the cattle in turn produced

[45] Purseglove, *Taming the Flood*, pp. 58-59.
[46] McManners, *Death and the Enlightenment*, p. 89.
[47] Bloch, *French Rural History*, p. 212.
[48] See Arthur Young, *Political Essays concerning the present state of the British Empire* (London: W. Strahan, T. Cadell, 1772), pp. 117, 130-32.

more milk which reduced waterborne disease and gave local children a much improved chance of surviving their first decade, the period in which mortality rates were highest. That the English model was dominant would appear to be proven by the fact that in 1800, at the height of hostilities between Britain and France, the revolutionary Directoire commissioned an eighteen-volume translation of selected agronomic works by Young, published under the title, *Le Cultivateur anglois*. Their author was suitably triumphalist in 1799, writing on the ubiquitous benefits brought to Lincolnshire by the drainage and reclamation of its fens: by this act the local population could boast "health improved, morals corrected and the community enriched".[49]

There is, of course, a more ambivalent post-script to this singularly progressive vision of eighteenth-century wetland "improvement". After all, it is not with impunity that man modifies such fine, natural "autoregulatory" ecosystems. Two points seem most salient in conclusion. Firstly, the technical advances which had initially facilitated wetland draining and enclosure were the prelude to yet greater technical advances in land management which ultimately deterred farmers from reclaiming marsh and fen; instead they opted for less expensive, less labour-intensive means of increasing productivity, such as buying into the bourgeoning nineteenth-century fertilizer industry. Combined with more frequent, cheaper foreign imports of foodstuffs, this second wave of the agricultural revolution left a lot of reclaimed wetland derelict, although it took a long time, if ever, to return to its former state of natural equilibrium. Secondly, and conversely, industrialization, with all its concomitant political and economic crises, far from sounding the death knell of peat farming, actually intensified interest in it as a low-grade fossil fuel source. By 1869 France was actually exporting 321,000 tonnes of peat a year at a price of more than ten francs per tonne. Peat farming remained a persistent, albeit embattled, staple of rural life in certain regions of France into the twentieth century, to the extent that the right-wing blood-and-soil novelist, Alphonse de Chateaubriant chose to set his extremely successful 1923 work, *La Brière*, among a community of peat-cutters in the Loire. Yet, more happily, the twentieth century also saw towards its end a belated recognition that wetlands were to be valued and conserved as unique natural wildernesses – ironically in

[49] Young, *General View of the Agriculture of the County of Lincoln*, p. 246.

the same way that the late eighteenth century had aestheticized and valorized the once dreaded wastes of mountain landscapes.

SECTION 2

EARLY TO MID-NINETEENTH CENTURY

"Nous avons enlacé le globe de nos réseaux…": Spatial Structure in Saint-Simonian Poetics

Greg Kerr

Abstract: *The early nineteenth century saw major developments in civil engineering and in cartography in France offering a new empirical relation to geographical space both within the French nation and further afield. These developments also offered a model for the elaboration of utopian political spaces which could dissolve the traditional opposition between nature and technology, a model that the Saint-Simonians in particular were keen to adopt. This chapter, then, explores a neglected source of inspiration for the Saint-Simonian reconfiguration of (largely urban) political space, through an increasingly abstract symbolism in the figurative use of telluric references and terms, developing a "panoramic" poetry.*

With the recording of new topographical information at the beginning of the nineteenth century, cartographers were able to produce an authoritative new map depicting France as a unified spatial entity. As Antoine Picon has shown, the task of developing this newly-defined territory fell to the modern corps of civil engineers.[1] During the Restoration and the July Monarchy, some of the most notable advocates of the development of the landscape were the utopianist Saint-Simonians, many of whom were engineers by training. Their project of social regeneration centred on a vast programme of public works,

[1] Antoine Picon, *French Architects and Engineers in the Age of the Enlightenment*, trans. by Martin Thom (Cambridge: Cambridge University Press, 1992), p. 101.

rural development and transport infrastructures which were intended to integrate human activities into the landscape and act as instruments for social unification, through a dynamic reconfiguration of the built and natural environments. The movement's initiator Claude-Henri de Saint-Simon (1760-1825) had from the outset demonstrated a preoccupation with the contemporary evolution of the disciplines of mechanics and hydraulics, and in the early 1800s he participated in a number of canal projects in the Netherlands and Spain; canals, railways and other networks were essential to his project of universal association, and they became recurrent figures of Saint-Simonian doctrine. Following Saint-Simon's death, and the eventual assumption of leadership of the movement by a self-styled "father", Barthélémy-Prosper Enfantin, the ideal of a functional organization of society remained a principal concern of Saint-Simonianism, and its adherents made major contributions to the establishment of the French railway network and the early plans for the Suez Canal. Thus Enfantin was able at the end of his life to claim: "Nous avons enlacé le globe de nos réseaux...".[2] Saint-Simonianism attributed crucial significance to the emergent understanding of the landscape as a complex spatial entity; in the text "Exposition du Système de la Méditerranée", for instance, the engineer Michel Chevalier envisages the mass construction of public works and transport infrastructures in terms of a dynamic reconfiguration of geographical space:

> Or quand il sera possible de métamorphoser Rouen et le Havre en faubourgs de Paris, quand il sera aisé d'aller non pas un à un, deux à deux, mais en nombreuses caravanes, de Paris à Petersbourg en moitié moins de temps que la masse des voyageurs n'en met habituellement à franchir l'intervalle de Paris à Marseille, quand un voyageur, parti du Havre de grand matin, pourra venir déjeûner à Paris, dîner à Lyon et rejoindre le soir même à Toulon le bateau à vapeur d'Alger ou d'Alexandrie ; quand Vienne et Berlin seront beaucoup plus voisins de Paris, qu'aujourd'hui Bordeaux, et que relativement à Paris Constantinople sera tout au plus à la distance actuelle de Brest, de ce jour un immense changement sera survenu dans la constitution du monde ; de ce jour ce qui maintenant est une vaste nation, sera une province de moyenne taille.[33]

[2] Cited in Gaston Pinet, *Écrivains et penseurs polytechniciens* (Paris: Paul Ollendorff, 1898), pp. 165-66.

[3] Michel Chevalier, "Exposition du Système de la Méditerranée: Politique nouvelle", *Le Globe*, 12 February 1832; repr. in Chevalier, Michel and others, *Religion saint-simonienne. Politique industrielle et système de la méditerranée* (Paris: rue Monsigny, Imprimerie d'Éverat, 1832), pp. 132-33.

The "spatial" idiosyncracies of this passage – visible in its animated transposition of exotic cities in the place of French regional towns – are suggestive of how Saint-Simonianism more generally manifests a certain interplay between utopian and poetic discourse. In this article, I will argue that the insights which inform such a project of geographical spatial appropriation simultaneously participate in the exploration of new poetic potentialities in more "literary" texts by Saint-Simonian authors such as Chevalier and the playwright Charles Duveyrier.

While Saint-Simonian propaganda had invested heavily in specialist technical literature throughout the 1820s, promoting in particular in the newspapers *Le Globe* and *Le Producteur* the civil and industrial benefits of such subjects as mechanical and civil engineering, under Enfantin the group soon became desirous of a less didactically structured and more generally *sensually persuasive* presentation of its programme for development. Enfantin's name is synonymous with the group's cultish retreat to a house belonging to him at Ménilmontant in Paris in 1832. There, Enfantin preached the "rehabilitation of matter", that is, a sensual reinvigoration of all those aspects of culture and society that had, in his estimation, fallen under the influence of Christian asceticism. One of the principal objectives of the retreat was the writing of the *Livre nouveau des Saint-Simoniens*, a manuscript intended as a prophetic synthesis of human knowledge. The work broaches a vast number of themes, reflected in its successive discussions of the liberation of woman, the potential social applications of electricity, mathematics, stereotomy and physiology, and the histories of language and literature. In the manuscript's record of a conversation between Enfantin and Chevalier, the former explains his desire to develop a more sensually appealing scientific model, or "science attrayante". In order to appreciate properly the benefits of this science, Enfantin says: "Il faut […] que nous entrions dans le sens des imaginaires".[4] The *modus operandi* of the "science attrayante" can be observed in the behaviour of select individuals:

> Or, il y a des hommes qui ont la propriété de mettre ainsi leur vie en dehors, dont l'existence se passe à combiner des modifications pareilles d'un objet, d'un être, sur

[4] Émile Barrault and others, *Le Livre nouveau des Saint-Simoniens*, ed. by Philippe Régnier (Tusson: Éditions du Lérot, 1991), p. 184.

> un autre être. Ils mettent toute leur vie entre des couples qu'ils unissent ainsi, qu'ils font couples. Ils sont sans cesse à voir des rapports entre des actes et des faits ou des idées que le plus grand nombre ne voit pas liés.
>
> [...] Il ne se passe rien en quoi ils ne cherchent un symbole. [...] Tout cela, c'est de la déraison, mais tu vois comment, dans le langage de ces hommes, il y a poësie. Tout pour eux est animé. Ils font de la mythologie perpétuellement, et tu vas voir ce qu'il y a de social dans cette vue d'animation générale.[5]

In his description of these inspired characters, it is noticeable that Enfantin selects a certain number of poetic qualities; amongst them the subject's communion with nature and an awareness of analogy and symbolism. Michel Chevalier goes on to agree with Enfantin that: "[...] dans le langage de ces hommes qui inventent des arts, il y a étonnamment de vie. Ils animent tout, pierres, métaux, engins, plantes [...]".[6] In Chevalier's view, the most enterprising minds also possess a peculiarly dynamic linguistic ability. Like Enfantin, Chevalier suggests that the vibrancy of their language enables their absorption into their material surroundings; these individuals are thus able to imagine new configurations of objects and environments.

Such vatic utterances by Enfantin and his disciples in the *Livre nouveau* inform a range of rhapsodic prose tracts and poems composed by prominent Saint-Simonians such as Chevalier and Duveyrier. For instance, Duveyrier's own prose poem dedicated to Enfantin, "Au Père!", presents the leader of the Saint-Simonians as a messianic figure whose ascension to glory is imminent. The speaker of the poem portrays himself as a precursor figure who literally clears the way for Enfantin:

> J'ai fait sonder les mers du gigantesque archipel. J'ai rassemblé comme une nouvelle nation d'Anglais contre les montagnes de la Chine et je leur ai donné le désir de franchir ces montagnes.
> [...]
> J'ai fait éclater de merveilleux spectacles à la face de ma terre. J'ai brisé de mon souffle les tempêtes qui rasaient le sol comme des lunes de malheur. J'ai pressé les mamelles des montagnes et j'en ai fait sortir leur lait de feu. J'ai souri en voyant les abîmes, comme des mâchoires de serpents, darder leurs flots dans l'espace, et j'ai fait glisser sur ces flots des villes armées aussi sûrement que sur la glace un patineur. Aux entrailles de la terre ferme, j'ai fait plonger l'homme comme un plongeur, et je l'ai fait voler au haut des nuées, vrai vautour! J'ai bâti des palais et des temples, des

[5] Barrault and others, *Le Livre nouveau*, p. 185
[6] Barrault and others, *Le Livre nouveau*, p. 186.

> ponts plus longs que des chaussées, et de fortes machines dont l'âme est de vapeur, les muscles d'acier, les flancs de fonte, et qui marchent seules.[7]

Here, the prospect of Enfantin's ascension – and thereby, the realization of the Saint-Simonian utopia – inspires the speaker to transform nature. This might be described as a sort of lyrical country planning according to which the earth is seen as a body whose udders can be milked and whose entrails can be explored. The introduction of numerous verbs of movement and multiple nouns into the phrase and the speaker's imagined transformation of earth, sea and sky produces pleasurable new configurations of sensation, reflected in the organic motifs of "mamelle" and "lait de feu". However, the poetic gaze never lingers to internalize what is glimpsed, as is seen in the suddenness of the "vrai vautour" and the prolonged description of the quasi-organic machine which ends with the abrupt cadence "et qui marchent seules". Elsewhere, Duveyrier's treatment of even the most minor details of the Saint-Simonians' programme of *aménagements* is related in highly expressive terms:

> Quelles fêtes que celles qui dresseront leurs tentes sur les flancs boisés des montagnes des Cauteretz et des deux Bagnères pour l'application des bains chauds aux douleurs du peuple, pour l'ouverture des plantations des Pyrénées, pour l'introduction de la vigne aux roches du Roussillon et le retournement et le labour des terres grasses et fraîches des plaines de Tarbes, des mamelons de fougères dorées au pays des Basques![8]

This ecstatic organicizing presentation of public works and routine agricultural reforms is continued later in the same text when he writes that: "[…] la terre nourricière tressaille du désir de nourrir et réjouir les misérables. La terre les appelle à fouiller ses entrailles […]".[9] In a discussion from the *Livre nouveau* Duveyrier complained that contemporary poetry exalted its objects, and that its contemplative tone placed barriers in front of sensual immediacy:

---

[7] Charles Duveyrier, "Au Père", in Barrault, Émile and others, *Le Livre nouveau des Saint-Simoniens*, pp. 210-22 (p. 211 and p. 219).

[8] Charles Duveyrier, "Travaux publics. – Fêtes.", in *Le Globe*, 16 April 1832; repr. in Chevalier, Michel and others, *Politique industrielle et Système de la méditerranée* , pp. 67-74, (p. 71).

[9] Duveyrier, "Travaux publics. – Fêtes." p. 73.

> Notre poésie, soit vers, soit prose, a plutôt quintessencié la matière qu'elle ne l'a énergiquement reproduite dans sa plénitude. [...] Le monde, pour les poètes, a été jusqu'à présent un mannequin revêtu d'un manteau magnifique et diapré de broderies éblouissantes; ils l'ont fait poser, et ils l'ont peint, mais immobile, mais inerte, mais froid, et eux-mêmes pour le peindre, ils se sont mis en manchettes! C'est qu'avec leurs habitudes chrétiennes de méditation, ils ont commencé par l'examiner, et à force de l'étudier, ils n'ont plus trouvé, au moment de le décrire, de fraîcheur d'impressions, de naïveté d'enchantement, d'élans d'inspiration! Ils ont voulu voir ce qu'il y avait dedans et ils ont manqué de passion pour le dehors. Il n'y a que notre foi qui puisse nous remettre au coeur, plus ardent qu'il ne fut jamais, l'enthousiasme de la nature ; et pour nous, il ne s'agit plus de l'adorer platoniquement, de célébrer sa régularité, son harmonie, et toutes ses perfections intimes, mais de nous livrer franchement à notre amour pour les beautés dont elle enivre nos sens. C'est alors que nous trouverons une langue vive, étincelante, neuve![10]

Duveyrier insists on sensual contact as a point of departure for poetic practice; he opines that by moving toward a form of expression constituted by the contact of the senses with the material world, the poet will sense an empathy with his surroundings that is unadulterated by Romantic melancholy and discover new sources of linguistic dynamism. Elsewhere, he writes:

> En vérité, lorsque je relis ces descriptions fameuses de toutes les merveilles du Nouveau-Monde, à voir le style symétrique, étudié, compassé avec lequel, par exemple, on décrit une des ces vastes forêts vierges où les arbres se pressent les uns les autres, ceux-ci tombant de vétusté et gisant à terre, ceux-là se heurtant, se balançant, fracassés par le vent, étalant enfin toute la verdeur d'une végétation vigoureuse, désordonnée, sauvage, je pourrais croire qu'elles sont alignées au cordeau comme le bois de Boulogne. C'est que l'orthodoxie de la phrase chrétienne subsiste: presque toute notre prose poétique y est entrée.[11]

Duveyrier calls for a rejuvenation of poetic language and a type of poetic prose that would supposedly restore an immediacy of contact with the Earth and more authentically convey the chaos of sense impressions. This concern with conveying immediate and uninterrupted sensual and visual stimulation registers in the massively inclusive poetic gaze adopted in his poems. In addition to the striking thematic impact of the incorporation and dynamic configuration of organic and artificial motifs in the text, Duveyrier's poetry demonstrates the markedly formal resonance of the latter through their potential to dynamize text

[10] Barrault and others, *Le Livre nouveau*, p. 123.
[11] Barrault and others, *Le Livre nouveau*, p. 124.

structure. A case in point is his lengthy prose poem "La Ville nouvelle ou le Paris des Saint-Simoniens", wherein rural idyllic images intrude into modern urban features, as shown in the following citation:

> Le bras droit de la bien-aimée de ma ville est tourné vers les coupoles et les dômes industriels, et sa main repose sur une sphère au sommet de cristal, à la surface enluminée du vert tendre des jeunes gazons, du jaune argenté des blés mûrs, et de toutes les nuances vives que les belles campagnes épanouissent sous les premiers baisers du matin.[12]

Here, the collision of pastoral imagery with "dômes industriels" gives rise to a re-organization of experience, and a sensual pleasure conveyed in the "nuances vives" and the "premiers baisers du matin". Such richly contrastive images are already inscribed in the topography of this imagined Paris that the gaze is never permitted to settle on and internalize what is glimpsed; Duveyrier seems to suggest that any meditative intervention would disturb the sensual pleasure of vision. Massive urban development does not give rise to the sort of melancholy and nostalgia later evoked by Baudelaire. Instead, such rapid change affords surprising visual contrasts and stimulates new configurations of sensation, whose pleasureable effects on the speaker nourish his empathy with the changing environment. To this end, the collision of "dômes industriels" – itself a clash suggestive of Renaissance architecture and nineteenth-century industry – with pastoral imagery creates an atmosphere of sensual surprise, giving rise to the linguistic fertility that is indicated by the "nuances vives".

Those readers who may have expected an impersonal Benthamite plan for reform of the city or a technical article of the calibre of many of the pieces published in *Le Globe* were to be surprised by the tenor of Duveyrier's piece. In the text, the angular and geometric figures which readers might have anticipated are displaced by organic and rounded motifs, including winding streets, curves and wheels: "Les places circulaires n'y sont pas plantées de quinconces régulièrement serrés et étouffés; des bouquets d'arbres s'élèvent çà et là comme les touffes d'herbes dans la campagne."[13] The buildings of the new city "s'élèvent en formes arrondies et bossueuses" while the architectural descriptions for the neighbourhood in question make fre-

[12] Charles Duveyrier, "La Ville nouvelle ou le Paris des Saint-Simoniens", in Barrault, Émile and others, *Le Livre nouveau*, pp. 222-36, (p. 233).
[13] Duveyrier, "La Ville nouvelle", p. 231.

quent reference to vegetation ("des champs de plantes grasses"... "des forêts de minces bambous"). Similarly, potentially sterile representations such as that of a timber yard may be expressed in exotic and sensual terms, for example, "les troncs des bois durcis dans les eaux tièdes de la Gambie et du fleuve des Amazones sont coupés par tranches comme les chairs d'un fruit fondant".[14] Within this shifting environment, a vast range of objects compete for the poet's gaze. These are glimpsed only in isolation, as in the following segments of phrases, framed by commas: "Des milliers de candélabres, *groupés en guirlandes autour des places*, *ou soutenus dans les airs sur des trépieds de cariatides*, prolongent dans toute la droiture de ma ville", "colonnes d'herbes géantes, *des grappes de fruits et de fleurs saillent des intervalles*; et ces sphères entassées", and "[a]utour de son vaste corps, jusqu'à sa ceinture, montent en spirale, *à travers les vitraux*, des galeries qui s'échelonnent".[15]

Each of the syntactical groupings or "interludes" (shown in emphasis here) are framed by punctuation, conveying an isolated glimpse. Lexically, the second of these examples draws attention to its own visual distinctiveness by the phrase "saillent des intervalles", suggesting a graphic immediacy of effect. Meanwhile, the third example presents the reader with a figure of visual fragmentation. The tensions generated by the new configuration of urban and rural features are also active on a stylistic level in the poem. Competition for the poetic gaze occurs, prompting a spontaneous borrowing from diverse registers. The text's panoramic ambit also encompasses tropes from fantasy literature ("les jardins aux fruits de neige et de glace"[16]). It is moreover possible to identify tensions internal to these tropes, as in the segment, "les monuments semblent descendre d'une grotte invisible, comme les palais de larmes du creux des montagnes, ou monter au ciel en légers cristaux".[17] Here, the gaze is undecided between a biblical "vale of tears" motif and a description of caves, mountains and crystals in the fantastic mode.

In "La Ville nouvelle" Duveyrier seeks to convey the poet's assimilation into an aggregate of sensation. To this end, one of the most emblematic figures of this poem (and of Chevalier's "Le Tem-

[14] Duveyrier, "La Ville nouvelle", p. 230.
[15] Duveyrier, "La Ville nouvelle", pp. 232-3.
[16] Duveyrier, "La Ville nouvelle", p. 228.
[17] Duveyrier, "La Ville nouvelle", p. 230.

ple", which will be analysed presently) is the solar plexus. Composed of a network of nerve endings, the solar plexus is promoted since it presents a figure of immediate, delocalized sensation. Duveyrier is keen to emphasize the interplay of the internal vitality of the individual and its external milieu, hinting that poetic subjectivity emerges at the point of contact between the subject and a complex and continually changing material environment. Consequently, his poem privileges qualities of extroversion and sensual immediacy over those of inward-looking contemplation and reflection, as in the persistent exclamatory utterances which litter the text ("Paris!", "Parthénon!", "Alhambra!", "Terre!"). In other instances, fresh sensual analogies are evoked:

> Ce sera comme une corbeille de fleurs et de fruits, aux formes suaves, aux couleurs tendres; de larges pelouses comme des feuilles les sépareront et fourmilleront de troupes d'enfants comme de grappes d'abeilles.
> […]
> Les rues sont sinueuses comme des anneaux qui s'entrelacent. Les murs sont couchés à terre, fermes et gonflés comme le turban d'un pacha, ou suspendus en l'air transparents et légers en des tresses de roseaux.
> Il s'élève du sol des colonnades et des voûtes qui sont semblables à des champs de plantes grasses dont les larges feuilles s'unissent en arceaux massifs, ou à des forêts de minces bambous au sommet desquels reposent des cloches, comme les fleurs sur leur [*sic*] tiges.[18]

Here colour, texture and shape are augmented, setting the objects viewed in a new light and reflecting a novel organization of experience. Moreover, exotic motifs such as the "plantes grasses" and "minces bambous" are awarded equal prominence to those of urban reconfiguration since they similarly imply a stimulation of the senses.

The massively comprehensive aesthetic vision implied in Duveyrier's vision is also a feature of the poetry of Michel Chevalier, in particular his 1833 poem entitled "Le Temple".[19] The poem describes an enormous temple which combines multiple architectural and organic references. On a thematic level, the text presents an alliance of technological advance with sensual excitement and exoticism. Its enumeration of a telegraph pole, lightning conductor, lighthouse, gaslights and other technical additions to the edifice artificially stimulate

---

[18] Duveyrier, "La Ville nouvelle", pp. 224, 231.

[19] Michel Chevalier, "Le Temple", in Barrault, Émile and others, *Le Livre nouveau,* pp. 237-43.

the senses and alter the poet-observer's cognitive processes, temporarily overwhelming him. This registers in the "electrification" of his language, as may be seen in the recurrent exclamations scattered through the poem. In the interior of the temple the poet glimpses steppes and savannahs, coconut trees and the Imperial Canal of China. Elsewhere, sensual stimuli are foregrounded via the inclusion of exotic motifs, notably in stanza Sixteen; here the encounter with Arabs, Chinese, Malays and Tartars is accompanied by coffee, tea, perfumes and feasts.

On a formal level, the alternations of line and stanza length suggest that the mode of the poet-observer's apprehension of this dynamic configuration of space is primarily fragmentary. One of the most metrically uniform passages of the poem occurs between stanzas Six and Eight, where line lengths range between seven, eight and nine syllables. This corresponds to the poet's identification of some of the more easily discernible features of the exterior of the structure; these include minarets, the telegraph pole, a lighthouse, towers and pyramids. The restricted line length suggests a pattern of succinct glimpses of each of these features, a point which would appear to be confirmed on a thematic level by the description of "le phare/ propice au navigateur".[20] By contrast, line lengths in stanza Nine are more expansive, as the poet's gaze penetrates the part of the structure which lies below ground. Unexpectedly, the movement inside and underground does not correspond to spatial economy or to a narrowing of viewpoint. Instead a wealth of perspectives is opened up as the poet's gaze reveals complex spatial deployments within the edifice's core; this is suggested lexically in this stanza by the "labyrinthes", "carrefours", and the "entrailles" which are "disposées avec un art infini".[21] Meanwhile, the rhythmical potential of alternate variations in line length is explored in stanza Twelve, notably in the following passage:

> […] les jets d'eau
> qui rafraîchissent les avenues et les portiques,
> les parvis et les voûtes;
> par les pierres et les blocs qui y sont suspendus
> ou qui le parsèment,
> par les cristaux taillés et colorés […].[22]

---

[20] Chevalier, "Le Temple", p. 238.
[21] Chevalier, "Le Temple", p. 238.
[22] Chevalier, "Le Temple", p. 239.

In this stanza, the rhythmical effect is underscored by the anaphora of the prepositions "par" and "de" throughout. In normal circumstances a preposition serves to indicate the temporal, spatial or logical relationship of its object to the rest of the sentence. However, with the repetition of prepositional phrases in this stanza the link to the clause "Le soleil y vit…" in the preceding stanza is no longer immediate. What results is a kind of parataxis *par contre-coup*, according to which the accumulation of prepositions actually isolates each phrase from the main clause, with the effect that successive images are juxtaposed. As the poet-observer experiences it, the world of the temple seems to be governed by contingency; however the act of repetition achieves a mesmerizing totality of effect which envelops him in the scene.

Chevalier's poem also suggests an intriguing formal receptiveness to contemporary developments in architecture and civil engineering. In *Le Livre nouveau*, Enfantin is attentive to the potential applications of contemporary developments in the science of elasticity, suggesting that future architectural designs would place less importance on adherence to principles of spatial form and composition inherited from Greek and Roman architecture and more recent neoclassicism. The latter, he claimed, were responsible for "despotisme contre la matière", resulting from a denial of the material substance and environmental situation of built structures.[23] As we have seen, Chevalier's text is formally resonant with similar concerns to those of Enfantin, through its expansion and contraction of the verse line according to the enumeration and accumulation of images by the poetic gaze. Such "elasticity" filters also into the metric structure of the poem, notably through the frequent deployment of imparisyllabic line lengths of seven and nine syllables and eleven and thirteen syllables. This formal peculiarity reflects a more flexible attitude to the conventional metric forms of the Alexandrine and the octosyllable. Together with the thematic incursion of pastoral motifs into urban scenes, and of that of the vegetal into the architectural, these formal and thematic aspects of the text combine to suggest the potential for new poetic forms to emerge organically around the armature of older variants. On this point, Chevalier's temple is a fundamentally styleless composite, but at the same time a dynamic synthesis of all possible architectural

[23] Barrault and others, *Le Livre nouveau des Saint-Simoniens*, p. 176.

styles; this much is suggested by the diverse architectural styles conjured up by pyramids, minarets, telegraph poles, by the pointed emphasis on the use of steel, copper, bronze and cast iron for the roof of the building and by the extremely diverse types of stone evoked in the following fragmentary verse:

> Cascades couvrant de leur écume les blocs de la
> Finlande, le granit de l'Altaï, le porphyre du Caucase,
> le marbre des Pyrénées, le calcaire des Alpes, les
> basaltes des Cordilières, les monceaux de fer de
> l'Afrique. [24]

A similar paradigm of amalgam seems to be implied when the speaker remarks of the temple that "toutes les végétations s'y unissent".[25] From an alternative angle, amalgam is also a writing strategy manifest in Chevalier's liberal borrowing from diverse literary styles. Utilitarian images ("terrains chauffés par combustion souterraine") abut orientalist tropes ("mystérieux asiles du plaisir" [...] "puits entouré des sables du Sahara") while expressions belonging to Romantic poetry ("Là règnent les amours les plus vives et les plus / profondes, etc.") are thrown into relief alongside figures from a mystical idiom ("la vie femelle dans les nerfs du plexus solaire").[26] Treated in isolation these elements possess only stock value, but their juxtaposition gives rise to a stylistic tension that dynamizes the text and renders it difficult to characterize according to conventional categories. By consequence, the poem seems to incite the reader to develop a panoramic perspective similar to that of the spectator it describes. Chevalier's ideal reader is one who reads in a spontaneous, un-premeditated manner and seeks neither to brood over the poem's technical achievements nor its original insights. Like the spectator, such a reader would not focus on individual elements but would cultivate a more immediate, panoramic perspective on the text. Chevalier's measure of poetic skill does not lie in a masterful knowledge of prosodic convention nor in technical perfection of and for itself, but in the deployment of rhetorical devices and syntactical idiosyncrasies which create a sense of incompletion and anticipation.

---

[24] Chevalier, "Le Temple", p.242.
[25] Chevalier, "Le Temple", p.242.
[26] Chevalier, "Le Temple", pp. 241-42.

The rhapsodic charge of the texts under consideration here indicates that the Saint-Simonians sought to pursue their planned reconfiguration of the landscape not just in terms of a utilitarian "travail public" but as a global project of social transformation drawing on all the material, lyrical and verbal resources available to them. This ideological ambition, with its implicit dismissal of individual imaginative consciousness, goes some way to explaining the ostensibly scant appeal of such Saint-Simonian visions of the future as a source of inspiration for creative writers of the period, for in reality, the Saint-Simonians were attempting to discover new, more persuasive bases for the articulation of a discursive regime. Moreover, Saint-Simonianism as political force entered a period of decline from the late 1830s onwards, and a considerable number of one-time members of the group such as Chevalier and Duveyrier came to be reconciled to the dominant political power. Yet these individuals continued to pursue and realize the economic, industrial and other projects that had been their concern during their period as Saint-Simonians; their utopian projects were of course evacuated of the contestatory quality invested in them by militant Saint-Simonianism, and as such became indissociable from the operations of power in the Second Empire. Yet, by the same token, many such former Saint-Simonians were responsible for the newspaper, advertizing, credit, exhibition and transport infrastructures which underpinned the Second Empire, to the extent that what appeared as some of the most visionary elements of the prophetic texts they composed in the 1830s came to be defining features of the material and social topography of the post-1851 regime, leaving us with what Pascal Ory calls, "un saint-simonisme lénifié, diffus mais d'autant plus opératoire",[27] and confirming Enfantin's conviction in the final triumph of the movement's permeative influence.

In seeking to account for the broader legacy of these Saint-Simonian texts to literature, it may be appropriate then to shift attention towards the dispersal of expressive strategies of the type identified in this chapter throughout other contemporary discourses which respond to changes in the material environment of modernity. Since poetry, and more particularly, the evolution of poetic form, represents one of the most striking manifestations of response to such changes, it should be possible, in the context of a broader study, to determine how

[27] Pascal Ory, *L'Expo universelle* (Brussels: Complexe, 1989), p. 10.

texts such as these anticipate (and thereby serve to illuminate) later developments in nineteenth-century poetics.

Pierre Leroux and the *Circulus*: Soil, Socialism and Salvation in Nineteenth-Century France

Ceri Crossley

Abstract: *This chapter examines attempts by the mid-nineteenth-century French socialist Pierre Leroux to establish an alternative, reformist relationship between man and the earth, one that focussed in a disconcertingly literal way on the recycling of human excrement as an organic fertilizer. This proved to be an original but futile challenge to both the growing chemical fertilizer industry and capitalist models of wealth circulation.*

At its most basic the *circulus* refers to the natural processes of ingestion, digestion and excretion. In nineteenth-century France, however, it took on a wider significance. Pierre Larousse's *Grand dictionnaire universel* (1869) proposed the following definition: "Théorie socialiste qui fonde le droit de vivre sur la possibilité pour chaque homme de suffire à sa propre subsistance au moyen de son propre engrais utilisé par l'agriculture".[1] Pierre Leroux (1797-1871) is generally credited with developing the economic, political and social consequences of the idea.[2] A prolific writer, philosopher and left-wing activist he used

[1] The entry was largely composed of quotations from Leroux, although it drew attention to similarities between his views and those of the American economists Henry Charles Carey (1793-1879) and Erasmus Peshire Smith (1814-82). On the question of the foundation of the right to life the article took Leroux to task, pointing out that human waste had to be transformed before being able to sustain life: "il faut du temps, du travail et du capital". Pierre Larousse, *Grand dictionnaire universel du XIX siècle*, 17 vols (Paris: Larousse, 1869), IV, pp. 335-36.

[2] For an excellent anthology see Pierre Leroux, *À la source perdue du socialisme français*, ed. Bruno Viard (Paris: Desclée de Brouwer, 1997). I would like to record

the theory of the *circulus* to challenge capitalist notions of the accumulation and circulation of wealth. His ideas concerning the *circulus* were formed in the 1830s and 1840s but they made their greatest impact during the Second Empire.[3] After the coup d'état of 1851 Leroux fled France and sought refuge in Great Britain. Like Victor Hugo, but in much more straitened circumstances, he spent much of his exile in the Channel Islands, returning to France after the amnesty of 1859. In his eyes the *circulus* gave priority to agriculture over industry, reconnected town and country, and inscribed humankind more generally within the processes of the natural world. Ultimately, the *circulus* became a form of theodicy, disculpating God (or Nature) from responsibility for evil and injustice. Leroux also opposed the use of chemical fertilizers and, for this reason, he stands as a significant forerunner of the contemporary movement in organic farming.

Philosophically Leroux belongs among the group of ageing Romantic humanitarians who, confronted by the rise of atheistic scientific materialism during the Second Empire, nevertheless held fast to their conviction that some form of belief in God was essential for the moral life to flourish. He wrestled with the problem of evil and with the reality of suffering. He recalled Joseph de Maistre's view that there was not a single instant of duration when one living being was not being devoured by another. Life arose from sexual reproduction and was maintained by the killing, absorption and assimilation of other creatures:

> Manger, voilà la loi primitive, l'origine et la clef de tous les phénomènes. Les anciens l'ont bien compris, et leurs langues l'ont bien exprimé. *ESSE*, *être*, disent

---

my thanks to Professor Viard for his invaluable help in the preparation of this article. For accurate information on Leroux and his writings see http://www.amisdepierreleroux.org/index.htm. See also (S.) Alexandrian, *Le Socialisme romantique* (Paris: Seuil, 1979), pp. 243-75; Jack Bakunin, *Pierre Leroux and the Birth of Democratic Socialism (1797-1848)* (New York: Revisionist Press, 1976); Armelle Le Bras-Chopard, *De l'Egalité dans la différence: Le Socialisme de Pierre Leroux* (Paris: Fondation Nationale des Sciences Politiques, 1986); Jérôme Peignot, *Pierre Leroux: Inventeur du socialisme* (Paris: Klincksieck, 1988); Vincent Peillon, *Pierre Leroux et le socialisme républicain: une tradition philosophique* (Latresne: Le Bord de L'Eau, 2003).

[3] Leroux first published his ideas about the *circulus* in March 1846 in his journal, *La Revue sociale*. See David Owen Evans, *Le socialisme romantique: Pierre Leroux et ses contemporains* (Paris: Marcel Rivière, 1947), p. 247.

toutes ces langues; *ESSE*, *manger*, ajoutent-elles. Le même mot signifie à la fois *être* et *manger.* [4]

What was to be done? Leroux did not side with those who, since the time of Pythagoras, had been arguing the case in favour of vegetarianism. His approach was different. Rather than engage in principled revolt against the natural order of things he sought to impose upon it the transforming vision of the *circulus*. Leroux's theory bound together the human, the natural, and the divine. It offered an alternative to the discourse of mastery that characterized so much of nineteenth-century thought, to the discourse that represented humankind as a demi-god controlling nature, overcoming all resistance and requiring that matter submit to the dictates of mind. The idea of the *circulus* enabled the possible emergence of a new collaborative relationship between humans and their natural environment. At the most elementary level Leroux was interested in improving the quality of the soil, an ambition that was shared by a host of contemporary agronomists in France and elsewhere in Europe.[5] How could the soil be made more fertile? What new forms of manure could be employed? Leroux addressed these issues but, as we shall see, he went further and argued that the proper use of human excrement could effectively replace money and release humans from enslavement to the wage-based economy that was the mark of modern industrial societies.[6]

Although Leroux often returned to the subject of the *circulus*, for the purposes of the present discussion I shall concentrate on the book that provides the clearest and most detailed statement of his position: *Aux Etats de Jersey sur un moyen de quintupler pour ne pas dire plus, la production agricole du pays* (1853).[7] The text runs to 227 pages, the first 79 of which focus on Leroux's proposals for renewing the island's sewage system. He provides a breakdown of the costs in-

---

[4] Pierre Leroux, *La Grève de Samarez*, ed. Jean-Pierre Lacassagne, 2 vols (Paris: Klincksieck, 1979 [1863-65]), II, p. 429.

[5] For a contemporary discussion of how to define soil see Constant Prévost, "Terrain" in *Dictionnaire universel d'histoire naturelle*, ed. Charles d'Orbigny, 13 vols (Paris: Renard, 1846), XII, pp. 477-79.

[6] See the highly perceptive article by Dana Simmons, "Waste not, want, not: Excrement and economy in nineteenth-century France", *Representations*, 96 (2006), 73-98.

[7] Page references cited in the body of the text are to Pierre Leroux, *Aux Etats de Jersey, sur un moyen de quintupler pour ne pas dire plus, la production agricole du pays*, (London: Universal Library; Jersey: Nétré, 1853).

volved, lists the anticipated benefits and explores how the project could be financed. A key proposal involved pumping treated human sewage to the fields through a system a pipes. Jersey already had a drainage system but this dispersed human refuse and abattoir waste into the sea (p. 45). If only the waste could be filtered and distributed in the form of liquid manure then, Leroux claimed, the island's agricultural production could be massively increased. Indeed, Jersey had the potential to supplant France and Belgium when it came to supplying London with early vegetables, flowers and fresh fruit. His programme was ambitious since it envisaged that fresh water, sea water and liquid manure would all be made available throughout the island.

Leroux shared the enthusiasm of his contemporaries for improved hygiene, sanitation and public health. The introduction of new systems of waste disposal was held to mark the triumph of civilization.[8] In *Aux Etats de Jersey* Leroux commented on the situation in both Paris and London. He discussed various initiatives supported by Prince Albert, quoted at length from the *Report of the General Board of Health* and took note of the contribution made by Charles Kingsley.[9] He explained to his readers that his attention had first been drawn to the subject in 1834. During a visit to the then failing Fourierist agricultural colony at Condé sur Vesgre he suddenly realised that the fortunes of the troubled establishment could easily be turned round if only fresh supplies of manure could be found. Shortly afterwards, while visiting one of his sons who was studying at the agricultural college at Grignon,[10] his mind became focussed on the idea that every creature produced enough excrement to sustain its own existence. This led Leroux to conclude that increased agricultural production required

[8] See Donald Reid, *Paris Sewers and Sewermen: Realities and Representations*, (Cambridge MA: Harvard University Press, 1991); Martin Monestier, *Histoire et bizarreries sociales des excréments des origines à nos jours* (Paris: Le Cherche midi, 1997).

[9] For a discussion of the obsession with excretion common among the French middle classes see Alain Corbin, *Le Miasme et la jonquille* (Paris: Aubier-Montaigne, 1982), p. 169. For the situation in Britain see Christopher Hamlin, *Public Health and Social Justice in the Age of Chadwick* (Cambridge: Cambridge University Press, 1998). For questions of cultural impact see Jonathan Ribner, "The Thames and Sin in the Age of the Great Stink: Some Artistic and Literary Responses to a Victorian Environmental crisis", *British Art Journal*, 1 (2), 2000, 38-46.

[10] Presumably the Institution Royale Agronomique established in 1826 on the site of Mathieu de Dombasle's model farm.

a massive increase in the use of human excrement as fertilizer. He was well aware of the dangers posed by untreated sewerage and deplored the activities of the unscrupulous Montfaucon night-soilers who produced vast quantities of "poudrette" (p. 107). He noted that the risks for health were greater in France than in England on account of the fact that the French favoured cesspits and lacked the water closets and sewerage systems that were being installed in England (although vast quantities of waste were still in fact polluting the Thames). His aim was to "changer une source de maladie et de pestilence en une source de richesse matérielle" (p. 191). What particularly irked Leroux was that, when such ideas received attention in the public press, his views were either ignored or misrepresented.[11]

Leroux, however, was not simply concerned with the implementation of more efficient schemes for the disposal of waste. The greater part of *Aux États de Jersey* was given over to a series of lengthy appendices that contained autobiographical anecdotes, political polemics and explorations of the implications of the *circulus*. Starting from the proposition that all of society's major ills were caused by hunger Leroux wondered how food production could be made commensurate with the needs of a growing population. His answer was breathtakingly simple. Food shortages would cease once human manure were used as a fertilizer: "le problème du Prolétariat envisagé comme problème des subsistances [n'est] qu'un problème de chaise percée" (p. 193). Leroux was immensely proud of his grand idea but was saddened to discover that his political enemies trivialized and ridiculed his theories.[12] In reality he was only one of many who considered that more extensive use could be made of human manure. In France the idea was much debated and official trials were success-

[11] Leroux was angered by a series of articles by Victor Meunier that, in his opinion, unfairly represented his authentically French, socialist ideas as so many foreign, English innovations. Leroux enjoyed casting himself in the role of a prophet crying in the wilderness but he did claim to have exerted some influence over Auguste Bella, the first director of the Institution Royale Agronomique (p. 103). Leroux was fully aware of developments elsewhere in Europe, notably the Kennedy system employed in Ayrshire to distribute animal manure to the fields.

[12] A political adversary of Leroux ended his discussion of the *circulus* thus: "Quand on voit un homme doué de facultés remarquables à certains égards tomber dans de telles aberrations, il ne reste qu'à sourire ou à gémir". Alfred Sudre, *Histoire du communisme ou réfutation historique des utopies socialistes* (Paris: Lecou, 1849), p. 466.

fully organized in Paris in the mid-1850s.[13] There was general agreement that human manure was an excellent fertilizer. According to the agronomist Jean-Baptiste Boussingault, writing in 1851: "*Les déjections de l'homme* sont un des agents les plus actifs dont dispose le cultivateur. Dans les pays où l'industrie agricole est en progrès, ces déjections sont très-recherchées, et on n'épargne aucune peine pour se les procurer".[14] The mid-century discussion regarding the benefits of human manure can usefully be related to the wider debate that was sparked by the development of artificial fertilizers. Justus von Liebig played a decisive role in promoting the use of chemicals, publishing his hugely influential study, *Chemistry in its Application to Agriculture*, in 1840. Science appeared to have demonstrated that what depleted soils needed most was the renewal of their mineral content. Although in some of his later work Liebig revealed himself to be a staunch supporter of manure, his prestige in the 1840s led many to conclude that the future of agriculture lay in the correct application of chemicals rather than organic matter.[15]

In France the advocates of artificial fertilizers referred to by Leroux – Jean-Baptiste Dumas, Anselme Payen and Jean-Baptiste Boussingault – were clearly winning the argument against traditional farming methods. What makes Leroux's ideas worthy of attention is the way in which his arguments prefigured a number of the principles that subsequently came to underpin the organic movement's rejection of industrial farming methods. Organic cultivation, writes the historian Philip Conford, aims to establish a virtuous circle: "the return of wastes to the soil creates humus, which encourages healthy crops whose remains, properly composted, return to enrich the soil's humus content".[16] The founders of the organic movement insisted that the natural order could not be "flouted with impunity".[17] Leroux would have agreed. He bluntly rejected the claims made by the chemists: "Je

[13] The results are discussed by Louis Figuier in *L'Année scientifique et industrielle* (Paris: Hachette, 1858), pt. 2, pp. 131-35.
[14] J.-B. Boussingault, *Economie rurale considérée dans ses rapports avec la chimie, la physique et la météorologie*, 2 vols (Paris: Béchet Jeune, 1851), I, p. 791. See F. W. J. McCosh, *Boussingault: Chemist and Agriculturist* (Dordrecht: Reidel, 1984).
[15] See William H. Brock, *Justus von Liebig. The Chemical Gatekeeper* (Cambridge: Cambridge University Press, 1997), pp. 250-72.
[16] Philip Conford, *The Origins of the Organic Movement* (Edinburgh: Floris Books, 2001), p. 17.
[17] Conford, *Origins of the Organic Movement*, p. 16.

ne crois pas à la philosophie chimique de Liebig, et je la nie" (p. 159). He accepted that nitrogen-enriched manures could make a contribution but he repudiated any attempt to explain soil fertility exclusively in terms of the combination of chemicals. Liebig's wrongheaded intention had been to substitute "un procédé *artificiel* au procédé de la Nature" (p. 178). Leroux criticized the ideas of Jean Girardin and Alphonse du Breuil who claimed that plants grew best in a soil containing naturally occurring mineral salts mixed with an appropriate amount of humus. Equally in error, in his view, were those who followed Albrecht Thaer and urged the extensive use of animal as opposed to human manure. In Leroux's judgment, while professional scientists accurately analysed the components of healthy soil, they displayed a disturbing ignorance regarding the manner in which the soil's constituent elements actually combined to form an entity that possessed life. Leroux observed that humans ate bread, not a mixture of flour and salt. And, like bread, the soil was alive and life-giving. The very idea of spreading factory-produced phosphates over the fields was anathema to Leroux who held that Liebig's followers were doing real harm to the French countryside. To prove his point he put his own ideas into practice between 1845 and 1848 at the agricultural colony that he and a group of his supporters set up at Boussac in the Creuse as part of his project for a socialist printing works.

Twentieth-century proponents of organic farming often drew inspiration from communities in India and China that had traditionally returned human waste to the earth. Leroux was of the same opinion. He revered the Chinese as masters and praised them for having instigated a system of agriculture based on natural law. In support of his arguments he reproduced extracts from a series of articles by a certain Dr Yvan that bore the title, "Les Vespasiennes chinoises ou supériorité des agriculteurs chinois sur ceux d'Europe" (pp. 144-49).[18] In Leroux's eyes the cycle of nature was a process that resembled that which twentieth-century supporters of the organic movement came to call the wheel of life. A chain of solidarity linked the animal, vegeta-

[18] According to Leroux, Yvan had visited China in an official capacity as doctor to the French Ambassador T. de Lagrenée (1800-62) who had been sent by the government in 1843 on a two-year mission to China that centred on negotiating commercial treaties between the two countries. Yvan's articles had originally appeared in the journal, *La Feuille du village* in 1849. The editor of the review, Pierre Joigneaux, was a friend of Leroux.

ble and mineral kingdoms. Change and decay were inevitable aspects of the cycle of nature but death did not extinguish meaning because the movement of transformation was purposive and self-sustaining: "La Nature a établi un Circulus entre la Production et la Consommation. Nous ne créons rien, nous n'anéantissons rien; nous opérons des changements" (p. 89). If Nature's resources were managed in accordance with the principle of the *circulus* no one would need to go hungry. Regrettably, however, humans had broken with the order of nature. Modern industrial society fostered division, competition and separation. It disdained the values of the *circulus* which was nature's way of bringing together the disparate elements of creation. Industrial society, founded on an arrogant individualism and the profit motive, had lost the sense that humans should abide by the very same principles used by nature "pour relier les différents êtres entre eux" (p. 97).

Leroux argued that in future the exploitation of human waste should be handed over to public authorities interested in alleviating the condition of the poor and no longer entrusted to individuals driven by the desire for private gain. The central focus of Leroux's ire was English economic liberalism which he viewed as a corrosive and destructive force, sundering humans from nature and precipitating a general crisis of civilization. Again and again he denounced Malthus and his followers, accusing them of using an inadequate model of nature in order to lend legitimation to the cruel and heartless society created by industrialism.[19] He was appalled by what he saw on the streets of London: crime, prostitution, unemployed labourers touting for work, Malays callously left to starve on the streets. How fortunate then that the theory of the *circulus* gave the lie to the Malthusian proposition that human population growth inevitably outstripped a society's ability to feed all its members. Leroux wanted to reconstruct the social bond by reconnecting humans to the order of nature. In opposition to Malthusian notions of competition, production and consumption, as an antidote to the egoism of his age, Leroux put forward "l'idée Socialiste du *Circulus*" (p. 27). This foregrounded connectivity. He started from the premise that the processes of digestion and excretion cannot adequately be understood in terms of the ways in which an organism extracts nutriments from ingested food before expelling the residue as

[19] See *Malthus et les économistes* (1849) in which Leroux republished six articles that had already appeared in his *Revue sociale*.

so much useless waste. In his view that which we perceive as mere waste has real value within the greater chain of solidarity. If only his ideas had been adopted by the Provisional Government of 1848 then, mused Leroux, the violence and bloodshed of the June days might have been avoided (p. 220).

What needs to be stressed, however, is the extent that, for Leroux, the potential practical benefits arising from recycling human waste were part of a world vision founded upon the ideas of interdependence, reciprocity and solidarity. God did not intend that dead or waste matter should simply be discarded. In reality, "les cadavres et les détritus des différents êtres peuvent être assimilés à des produits préparés pour la subsistance d'autres êtres" (p. 10). Human manure should be returned to the earth in order to enrich the soil and aid the production of food. "Dans une infinité de cas," continued Leroux, "la vie entretient la vie par des produits qui, pour être utilisés, n'entraînent pas la cessation d'existence de ceux qui les donnent" (p. 10). Death was not absolutely necessary in order for life to continue. The products of excretion, far from being without worth, were intended to fulfil a positive role within the cycle of life. It was an error to understand the act of eating purely in terms of the satisfaction of appetite: "l'animal ne digère pas pour lui seul, mais, si je puis employer cette expression, digère pour préparer l'alimentation de la plante" (p. 94). Leroux challenged received opinion regarding the status of urine and excrement. He argued that it was incorrect to draw too sharp a distinction between excretion and secretion, compared urine with milk and regretted that he had not been able to write a study on the consumption of urine as he had once intended (p. 140). In a similar spirit he took on Bichat and the anatomists whom he accused of failing to grasp that the complex digestive system of animals was a consequence of nature's refusal to allow for consumption without production (p. 102). Leroux explained that when food passed through the alimentary canal something more complex than straightforward assimilation took place. Something new was actually added during the process (p. 162). Bichat had not grasped the relationship between the large intestine and the caecum (p. 192). Berzelius, on the other hand, received praise on the grounds that he had noted that something new was added during the passage of ingested matter through the intestines (p. 102).

Again and again Leroux reinforced this point that life was supported by a set of interlinked bodily functions. Animal waste that was returned to the earth enriched the soil. Cats and certain other carnivorous animals instinctively covered their excrement because they knew that it needed to be mixed with minerals and vegetable matter in order to become productive. In Leroux's mind the processes of digestion and excretion became a unifying metaphor, bringing all life forms together within a greater unity. He was alert to similarities and analogies. He seized upon some remarks made by the Swiss botanist Augustin Pyrame de Candolle who had discovered the presence of small lumps that resembled excreta on the root systems of certain plants.[20] For Leroux these lumps were more than waste matter. They had a definite role to play in the grand scheme of things: "Les végétaux, comme les animaux, digèrent pour autrui, tout en digérant pour eux-mêmes; ils reçoivent et donnent; ils rendent, d'une centaine à la terre par leurs excréments d'une certaine façon ce qu'ils lui empruntent" (p. 96). Leroux argued that it was the excrement discharged by one type of plant that made the soil fertile when another species grew there. In other words the efficacy of crop rotation did not depend, as scientists such as the Swiss chemist Théodore de Saussure had claimed, on new plants drawing their nourishment from decaying vegetable matter but on the operation of a natural law. Humans should follow the example set by the plants and animals and return their own excrement to the land.[21]

Leroux described as a "composé *vivant*" (p. 119) a soil that was genuinely healthy and fertile. He accepted that decaying vegetable matter produced humus but he contended that this, on its own, was

---

[20] Discussion of such matters was current among specialists. Augustin Pyrame de Candolle's ideas were examined by his son, Alphonse de Candolle, in his *Introduction à l'étude de la botanique* (Brussels: Meline, 1837), pp. 140-41.

[21] According to Leroux it was this guiding principle that lay behind Moses's injunction in Deutronomy 23. 13: "And thou shalt have a paddle upon thy weapon; and it shall be, when thou wilt ease thyself abroad, thou shalt dig therewith, and shalt turn back and cover that which cometh from thee". Michel Lévy, a leading public health reformer of the Second Empire agreed with the need to take extremely seriously the views expressed in the Bible: "les préceptes sanitaires de la Bible procèdent d'un système de préservation collective, non de quelques conjectures incohérentes", *Traité d'hygiène publique et privée* (Paris: Baillière, 1857), 1, p. 5. Lévy quotes the same lines from Deutronomy as Leroux and comments: "Ce précepte, que le soleil de l'Arabie rendait si urgent, est oublié aujourd'hui dans ces mêmes lieux où il a été dicté, jusque dans les villes, au grand détriment des populations" (p. 8).

not enough. Truly fertile soil was a combination of animal, vegetable and mineral elements. It became "un *composé*, un *mixte*, une substance à un certain degré vivante" (p. 118). Leroux proudly told his readers that during his exile in London he had successfully grown a crop of beans having first made soil by mixing his own faeces and urine with coal, ashes, brick and sand, actions that vindicated God's words when he told Ezekiel to bake barley cakes using "dung that cometh out of man" (Ezekiel 4. 12). The man who used his own faeces to make soil and grow food completed the circle of life and reintegrated humankind within the purposeful totality of living things. Were such practices to become generalized then there would be enough meat and good quality bread to support an increasing population (p. 21). It would be like a return to Eden. However, the implications of the theory of the *circulus* went beyond the eradication of hunger. The triumph of the *circulus* signalled the defeat of egoism in the name of solidarity: "Le *cercle naturel* remplace la *circulation* des économistes, qui n'en est qu'une odieuse contrefaçon" (p. 179). The *circulus* was "le moyen par lequel cette Nature unit tous les êtres" (p. 177). The *circulus* re-established "l'unité de la vie et la chaîne qui lie tous les êtres" (p. 176). It constituted a religious truth and inspired faith (p. 196). The Word was to be preached abroad and enacted in practical living (p. 196).

In his younger days Leroux had been a Saint-Simonian. Central to Saint-Simonian doctrine had been the desire to rehabilitate matter, to restore to the material universe the value that had traditionally been denied to it by the Christian tradition. In his writings on agriculture Leroux effectively extended this notion of rehabilitation to include faecal matter and bodily fluids: "la question véritable est de savoir si les excréments ont une valeur, et quelle est cette valeur" (p. 137). Humans had failed to understand the depth of meaning surrounding excretion:

> Voilà des siècles que les hommes satisfont tous les jours à une fonction naturelle sans la comprendre!
> Leurs déjections, après qu'elles sont sorties de leurs corps, leur sont inutiles et odieuses: ils disent donc: Ces déjections ne servent à rien.
> Ils ne sauraient s'en nourrir; au contraire, la Nature a arrangé tout de manière à ce qu'ils ne fussent pas tentés de le faire. Ils disent donc: Cela ne peut pas être une nourriture pour d'autres êtres.
> Leurs sens sont blessés, ils disent: Voilà une chose abominable. (p. 137)

These lines indicate the extent to which Leroux's project involved the moral rehabilitation of human waste (in addition to an assessment of its practical – and commercial – value as a fertilizer). From Leroux's socialist perspective the way in which excrement reconstituted the health of the soil was the supreme demonstration of the interdependence of all living entities:

> Si on vous disait que le rôle que vous faites jouer à la Nature est un rôle ignoble et indigne d'elle; que c'est *l'égoïsme* le plus grossier que vous imaginez être sa loi, tandis que sa loi est l'harmonie absolue entre *l'égoïsme* et le *dévouement*; qu'elle n'a pas créé un seul être pour lui-même, mais qu'elle les a créés tous les uns pour les autres, et a mis entre eux tous une solidarité réciproque! (p. 154)

When it came to transforming and using nature Leroux implied that humankind should act wisely and take account of the bonds that linked the microcosm and the macrocosm (p. 87). The error of the Malthusians had been to ignore the true message of nature; genuine social progress involved respecting natural law, uniting with the general movement of the cosmos. However, while this indicated that the future for France lay in agriculture rather than industry, Leroux was not by any means a Luddite. After all, his plans for the distribution of manure required the construction of complex systems of piping designed to run alongside railway lines. The whole was greater than the sum of the parts but the parts needed to combine with each other in order for the whole to thrive and prosper. Leroux offered an excremental parable: "presque partout les excréments des oiseaux et des autres animaux ont servi à *faire de la terre* et à nourrir les plantes, et, par les plantes, les animaux et l'homme" (p. 206). The removal of humans from the natural cycle led to dislocation and ultimately to sterility – hence the moral imperative to reintegrate human waste into the *circulus*. Healthy soil, as we have seen, was a composite that arose from a collaborative process. Were this not to be the case then the landscape would run the risk of being suffocated beneath an increasingly thick layer of human guano.

Leroux's argument offered an alternative to the Malthusian vision of nature. It also answered the Maistrian vision of life on earth as generalized violence, death and consumption. Leroux proposed something different: "l'alimentation des êtres par la vie des autres êtres" (p. 10). Here was a chain of consumption and production that involved giving and receiving. Leroux was convinced that his vision of nature lent support to the socialist cause. He explained that an individual plant, interested exclusively in its own survival, would soon perish. Selfishness decreased the chances of survival. To imagine that a plant selfishly drew water and minerals from the surrounding soil and then repaid its debts to the earth when its leaves finally fell to the ground was to betray a singular misunderstanding of the workings of nature. In reality, continued Leroux, the fallen leaves fertilized the soil for the wider benefit of other plants. The continuance of life on earth rested on similar complex processes of sharing and exchanging. Malthus had correctly recognized the infinite fertility of living things but he had not grasped the true character of natural law. Nature was debased and traduced when its processes were used in order to lend legitimacy to economic liberalism. It was quite wrong to draw an analogy between nature and a banker who was interested in profit and loss and expected to be repaid (p. 135). The operation of the *circulus* worked against the exploitation of the weak by the powerful. It was intrinsically anti-hierarchical in character.

From Leroux's socialist perspective the earth was the collective property of the living entities whose waste products sustained life on the planet: "Nous sommes tous, suivant moi, *fabricateurs de terre*, et je maintiens même que toute terre, et par conséquent toute propriété, a été ainsi fabriquée, soit par les animaux, soit par les hommes" (p. 193). Leroux wanted to link the individual's obligations with regard to his fellows to humankind's general duty towards the planet and he did this by interpreting excretion as a form of restitution. He described the human body as "un admirable laboratoire" (p. 11).[22] The processes of digestion, assimilation and excretion undertaken by the organs of the body accorded with the divine plan of creation.[23] Nothing was wasted since what was perceived to be waste, impure and unclean, was in fact essential for the continuance of life. It followed from this that, in Leroux's eyes, the death of a human being from starvation was not an injustice that could be blamed on God. The

[22] For an illustration of how the functioning of the digestive system could be represented as a model for a well organised society see Paul Gaubert, *Hygiène de la digestion avec quelques considérations nouvelles* (Paris: Au Dépôt de la Librairie, Rue Sainte-Anne), 1849, p. 121.

[23] See the comments on Leroux's theories made by Dominique Laporte in *History of Shit*, trans. by N. Benabid and R. el-Khory (Cambridge MA: MIT Press, 1993), pp. 128-32.

fault lay with a society that was not organized in accordance with the principles of the *circulus* (p. 11). In this manner Leroux's rehabilitation of excrement absolved Providence of responsibility for human suffering (p. 29). Waste matter underwent a metamorphosis as it transformed the soil into that which it was intended to be. Leroux epitomized the Romantic desire to redefine the relationship between infinity and the finite, time and eternity, heaven and earth, matter and spirit, the sacred and the profane. He believed that his contemporaries needed a new unifying faith and he attempted to construct it, blending humanitarianism with nationalism, the revolutionary idea with perfectibility, equality and solidarity with individual freedom and private property. Central to this project was the attempt to invest the present time and the material universe with a sacred dimension: "L'intelligence divine éclate dans toute la Nature, nous vivons pour ainsi dire en pleine Divinité. Subjectivement, objectivement, nous trouvons Dieu. Nous le sentons en nous, nous le découvrons dans chaque grain de poussière."[24] The *circulus* was the manifestation of life, the engine of progress. Its operation disrupted received definitions of spirit and matter, purity and impurity. It allowed the emergence of new definitions of labour, capital and consumption. The feelings of disgust, revulsion and shame engendered by the sight of waste did not tell the whole story. As if by magic, the *circulus* converted sterility into fertility, base matter into something of positive value. By attending to the soil and to the nature of its composition humans could learn important truths, not only about agriculture, but also about themselves and the organization of society.[25]

[24] *La Grève de Samarez*, II, p. 426.

[25] Leroux did not emphasize the health benefits of "natural" foods though, like many contemporaries, he was concerned by the adulteration of food. He was interested in animal welfare insofar as he objected to cows being kept permanently in stalls (p. 104) and viewed recent developments in stock-breeding with alarm (p. 142).

Mind as Ruin: Balzac's "Sarrasine" and the Archaeology of Self

Scott Sprenger

Abstract: *While earlier critical studies of Balzac link his work to archaeology in a literal way, this chapter contends that the writer configures narrative consciousness as a site on which a kind of "archéologie morale" is to be practised; that is, as the site of a construction of the self that has confronted or will confront catastrophe – natural or otherwise – only then to undertake an archaeological process of reconstruction to make sense of it. This archaeological interpretation of Balzac's fiction is illustrated by a close reading of the novella* "Sarrasine".

Freud speculated on the causes of modern psychopathologies by figuring the mind as an ancient city in ruins. He postulated that, like an archaeological site, the modern mind is structured in temporal layers and that forgotten or repressed events from the past can be reconstructed from fragmentary remains. In this new, archaeological figuration of the mind, Freud challenged the conventional Enlightenment conception of it as unitary, rational and master of its conscious will. Indeed, he exploded the traditional, rationalist view by speculating on the ego's irrational and largely unknown underside – the unconscious. In an extended passage in *Civilization and Its Discontents*, Freud invokes the archaeological layers of Rome's ancient cityscape to help readers imagine the internal layering of the conscious and unconscious portions of the modern psyche:

> [W]e have been inclined to take the [...] view, that in mental life nothing which has once been formed can perish – that everything is somehow preserved and that in suitable circumstances [...] it can once more be brought to light. Let us try to grasp what this assumption involves by taking an analogy from another field. We will choose as an example the history of the Eternal City. . . . If [the observer] knows enough – more than present day archaeology does – he may perhaps be able to trace out in the plan of the city the whole course of [the wall of Aurelian] and outline of the Roma Quadrata. [...] There is certainly not a little that is ancient still buried in the soil of the city or beneath its modern buildings. This is the manner in which the past is preserved in historical sites like Rome.
>
> Now let us, by a flight of imagination, suppose that Rome is not a human habitation but a psychical entity with a similarly long and copious past – an entity, that is to say, in which nothing that has come into existence will have passed away and all the earlier phases of development continue to exist alongside the latest one. . .[1]

The point of insisting on Freud's use of this literary "flight of imagination" to leap from the ruined city to the inner debris of the modern mind is not to criticize Freud for analogical thinking or to undercut psychoanalysis's value as a legitimate human science (although there are other reasons to do this), but to underscore the blurry frontier between literature and social "science". Literature has in the modern academy become the poor cousin of the social sciences, although many contemporary social sciences, including sociology, psychology and even psychoanalysis, originally emerged from literary observations and figurations. Freud was a voracious reader of literature and shamelessly lifted metaphors, analogies and mythical figures in developing his various models of the unconscious mind. Thus, while it has been common practice for several decades to apply social scientific methods such as psychoanalysis to literary criticism in order to lend it "scientific" prestige and sophistication, we propose to do the opposite: to show how a novelist such as Balzac used archaeological metaphors to imagine a new science of the mind that anticipates psychoanalysis while exposing its epistemological limits. Like Freud, Balzac was obsessed with archaeology and archaeological modes of narration. Also like Freud, Balzac discovered powerful heuristic potential in archaeology as he began to suspect that adult psychopathologies were secretly rooted in forgotten or repressed childhood events. One major

---

[1] Sigmund Freud, *Civilization and its Discontents*, trans. James Strachey (New York and London: Norton, 1961), pp. 16-17.

difference, however, is that while for Freud the unconscious is formed from the repression of erotic desire during the Oedipal stage of development, for Balzac the unconscious is formed from a set of traumatic historical events: the revolutions – 1789, but also 1830 – intended to modernize and "nationalize" French consciousness by displacing Old Regime/Catholic desires, sentiments and habits. For Balzac, modern French consciousness emerges in its fragmented, modern state as it crosses the historical and epistemic divide between old and new France.

The following interpretation of "Sarrasine" will show that while psychoanalysis has historically been one of the most popular approaches to interpreting Balzac's fiction, it may actually impede understanding of his theory of post-revolutionary consciousness, encouraging readers to misrecognize the Catholic content of its internal displacements. Modern desire, according to Balzac, is always underwritten by a residual religious desire, aspiration or mental habit, which requires new hermeneutic tools – an archaeology of consciousness or what Balzac called "moral archaeology" – in order to be perceived and properly understood.

The numerous references to archaeology in *La Comédie humaine* have not passed unnoticed by critics, even if, unlike Gautier, Mérimée or Flaubert, Balzac never represented an ancient civilization in his fiction. Balzac's interest in this discipline has been either noted or analysed in various critical works, such as Jeannine Guichardet's *Balzac, "archéologue de Paris"*. Yet contrary to these studies,[2] concerned either with the thematic and architectural dimensions of archaeology in Balzac (Guichardet, for example, exhaustively catalogues archaic or ruined edifices in Paris) or with the fragmentation and discontinuities of Balzac's writing as postmodernity *avant la lettre* (Barthes, Dällenbach), we propose to show Balzac's use of archaeology and archaeological metaphor in theorizing fragmented and layered consciousness,

[2] See Philippe Bruneau, "Balzac et archéologie", *Année balzacienne*, (1983), 15-50; Nicole Mozet, "La mission du romancier ou la place du modèle archéologique dans la formation de l'écriture balzacienne", *Année balzacienne*, (1985), 221-8; Boris Lyon-Caen, *Balzac et la comédie des signes*, Paris, Presses Universitaires de Vincennes, 2006, pp. 31-33; Jeannine Guichardet, *Balzac, "archéologue de Paris"* (Paris: CDU-SEDES, 1986).

of which the surface level is shaped by a deeper, involuntary layer of habit or emotion.[3]

Consider, for instance, the examples of archaic or fragmented consciousnesses that Balzac presents as being “en ruines”. Most prominent are, of course, nobles or provincials who remain attached to their archaic traditions although they live, often unwittingly, in modern environments. Balzac metaphorizes the ancient dimension of such characters by using words such as “antique”, “reliques”, “débris”, “fantômes”, or “mort-vivants”. The countess, Stéphanie de Vandières (in *Adieu*) whose collapse into madness follows traumatic separation from her pre-revolutionary world, or the *ultras* of *Cabinet des antiques*, considered by Balzac as “antiques” because they refuse to modernize, represent this type. Parallel, however, to characters explicitly labelled as “antiques” are others who claim to welcome modernity (republicans, scientists, philosophers, artists, businessman, etc.), yet who are nonetheless animated by residual habits and reflexes, analogous to a living fossil. Their otherness, repressed within the modern self and abandoned by the rationalist human sciences of the period, is the object that Balzac wishes to reconstruct through the new science of “archéologie morale”.[4]

Balzac understood that this interior moral reality was purely illusory from a rationalist or materialist point of view. But unlike the “official” scientists of his time, he was prepared to consider “real” that which others considered pure illusion or abstraction. Put differently, moral archaeology permitted Balzac to study human reality by adopting a sociological or anthropological approach, and thus to determine culturally-determined causes of behavioural effects. To the extent that many of Balzac’s contemporaries remained fixated on obsolete cultural or religious illusions, and to the extent that these fixations caused real (i.e. observable) psychopathologies, Balzac believed it necessary to invent a new human science capable of examining these “imaginary” phenomena from an analytical or scientific point of view.

The theme of repression, internal otherness, *de l’insu*, etc. evokes the unconscious. The Balzacian unconscious is not yet, how-

---

[3] See my article, “Balzac, archéologue de la conscience”, *Archéomanie: La mémoire en ruines*, ed. by Valérie-Angélique Deshoulières and Pascal Vacher (Clermont Ferrand: Presses, Universitaires Blaise Pascal, CRLMC, 2000), pp. 97-114.

[4] I extrapolate this term from the syntagm “archéologue moral” found in the opening pages of the novel, *Béatrix*.

ever, the Freudian unconscious, since for Balzac modern French consciousness is grounded in an historical rupture: the sudden break with feudalism and Catholicism realized in Republican modernity, rationalism, individualism and secularism. The "other-of-the-self" emerging from this rupture is thus a conglomeration of residual feelings, habits and memories that animate consciousness and behaviour but do not align with the rationalist concept of the self inherited from the Enlightenment and enacted by modern Republicanism. Balzac, like Freud, was probably atheist, considering the metaphysical substance of religion a simple illusion. As a moral archaeologist, however, he clearly recognized the residual influence of Old Regime customs on modern consciousness. It is this contradiction between the myths of modernity (individualism, rationalism, freedom from Catholicism, etc.) and the residual religious feelings hidden beneath that Balzac sought to expose.

In novels such as *Louis Lambert*, *Adieu*, *L'Auberge Rouge*, *La Recherche de l'absolu* or "Sarrasine", Balzac systematically uses archaeological metaphors to draw attention to the correspondence between psychical discontinuity in characters and the historico-cultural discontinuity of early nineteenth-century France. The metaphors signal an epistemic rupture between apparently conscious thoughts and behaviours at an adult age and archaic, involuntary habits from childhood (and/or a bygone era) that nonetheless determine thoughts and behaviours. Often, after a traumatic event, characters find themselves distanced from their affective or "moral" past, leading to symptoms of nostalgia or madness that Balzac codes metaphorically as archaeological debris in order to signal that the problem originates in the remote past and consequently eludes voluntary memory.

The influence of scientific archaeology can be found throughout Balzac's work, from his early theoretical treatises to his last novels. In *La Théorie de la démarche*, for example, the novelist compares his way of decrypting human behaviour to the Egyptian paleontology of Champollion, "[qui] a consumé sa vie à lire les hiéroglyphes [...] et nul n'a voulu donner la clef des hiéroglyphes perpétuels de la démarche humaine" (XII, 261). Balzac was also inspired by the naturalist Georges Cuvier, whose reconstructions of lost forms of animal life sparked the idea of reconstructing lost forms of human consciousness: "Il existe une anatomie comparée morale, comme une anatomie comparée physique. Pour l'âme, comme pour le corps, un détail mène logiquement à l'ensemble" (XII, 283). In the "Avant-Propos" of *La Comédie humaine*, Balzac presents himself as "l'archéologue du mobilier social [cherchant] à surprendre 'le sens caché' dans cet immense assemblage de figures, de passions, et d'événements" (I, 52). Likewise, in *Béatrix*, the narrator explicitly compares his function to that of the moral archaeologist, "[qui observe] les hommes au lieu d'observer les pierres" (II, 8). The opening pages of *La Recherche de l'absolu* offer the clearest exposition of the theoretical and methodological stakes of Balzac's moral archaeology, in his innovative analogy between the archaeological reconstruction of a civilization from its material remains and of the unknown, forgotten or repressed parts of an individual psyche:

> [L]a plupart des observateurs peuvent reconstruire les nations ou les individus dans toute la vérité de leurs habitudes, d'après les restes de leurs monuments publics ou par l'examen de leurs reliques domestiques. L'archéologie est à la nature sociale ce que l'anatomie comparée est à la nature organisée. Une mosaïque révèle toute une société, comme un squelette d'ichthyosaure sous-entend toute une création. De part et d'autre, tout se déduit, tout s'enchaîne. La cause fait deviner un effet, comme chaque effet permet de remonter à une cause. Le savant ressuscite ainsi jusqu'aux verrues des vieux âges. (X, 657)

Balzac's conception of modern consciousness as "stratified" ruins owes much to the rise of scientific archaeology in the early nineteenth century, but is also grounded in observation of the layers of culture generated by the rapid succession of French post-revolutionary regimes. He illustrates the process of fragmentation and internal repression that contemporary consciousnesses experienced in crossing the epochal threshold from Old Regime to modernity.

Balzac's archaeological theory of consciousness thus anticipates Freud's layered view of the psyche. Passionate about archaeology himself, Freud openly declared in *Constructions in Analysis* its importance in his invention of psychoanalytic methodology and narrative.[5] Freud and Balzac differ, however, in that the repression shaping the Balzacian unconscious is not (only) sexual or Oedipal; it is the repression of Catholic habits and aspiration by modernity and the forced passage of traditional consciousness into the post-revolutionary order that causes internal separation. To apply psychoanalysis to Balzac

[5] Sigmund Freud, "Constructions dans l'analyse", in *Résultats, idées, problèmes,* 6th ed., (Paris: PUF, 2005), p. 271.

without considering the historical dimensions of the formation of modern French consciousness is not only to misinterpret the stakes of Balzacian psychology but also to repress the crypto-religious dimension of modernity and modern consciousness that Balzac seeks to expose.

In "Sarrasine", for example, Balzac secretly undertakes an "excavation" of the eponymous character's consciousness. Numerous studies, including those inspired by Barthes's *S/Z* – which many consider the definitive reading of "Sarrasine" – have attempted to resolve the mysteries of the text without, however, managing to explain everything. Our hypothesis is that the narrative, textual and imagistic fragmentation characterizing the text is not gratuitous and does not necessarily correspond to a postmodern or poststructuralist point of view. Sarrasine undergoes a spiritual "catastrophe", emerging from it with his consciousness radically divided, with the archaic/mystified part embodied in Sarrasine and the anonymous narrator incarnating the modern/rational part. Sarrasine is born and raised in a pre-revolutionary context, receiving a religious education with the Jesuits, travelling to Rome to study art and eventually discovering the object of transcendent love (la Zambinella) who drives him crazy and whose absence of love "kills" him. As a child, Sarrasine is described as "sauvage" and "bizarre", a family outcast, resistant to the will of authority, including that of his father. The narrator, by contrast, is a debauched *mondain* in nineteenth-century Paris. He considers la Zambinella – without illusions and many years later – as an old man "en ruines", "un débris humain", "un spectre" or "un cadavre ambulant".

Barthes explains that the fragmentation of "Sarrasine" results from "l'effondrement catastrophique" caused by the castration of la Zambinella, which breaks the unity of the text, unleashing a proliferation of metonymies: "Le champ symbolique est occupé par un seul objet dont il tire son unité [...]. Cet objet est le corps humain. En somme la nouvelle représente [...] un effondrement généralisé des économies... Cet *effondrement catastrophique* prend toujours la même forme: celle d'une métonymie effrénée".[6]

Initially, Barthes's reading appears to be correct, revealing an undeniable epistemic separation between the narrator and his object

[6] Roland Barthes, *S/Z* (Paris: Éditions du Seuil, 1970), p. 221. Subsequent page references to Balzac's "Sarrasine" will be taken from this text, and will be given in brackets after the relevant quotation.

that renders his language, images and metaphors fragmentary. The narrator is, indeed, able to offer only a partial and incomplete idea of his object despite the excessively long – even obsessive – description of him. The old man is represented as a fragmented vestige from another era, but the key element that would give finality to the portrait remains missing. Barthes's insistence on the text's plurality and infinite signification from the object's undecidability thus seems plausible.

An "archaeological" reading of "Sarrasine", however, demonstrates that the rupture at the heart of the text is self-consciously strategic and has a specific narrative function. The fragmentation plays an important role in the archaeological reconstruction of the consciousness of *the narrator*'s psychical trauma and internal separation. The puzzle pieces that the narrator presents in the first half of the narrative (the description of the Lanty family, the fragmented condition of the old man, the picture of the Adonis, etc.) are residual effects of an affective catastrophe that the narrator underwent in his past. Barthes's insistence on the "infini du langage" would have us believe that there is no end to signification, forestalling any attempt to discover a method for reassembling the fragments of the narrator's lost illusion.

To identify the narrator's catastrophic experience (which he never actually mentions) and the correspondence between Sarrasine's catastrophe and the narrator's divided consciousness, we must reappraise the narrative, thematic and allegorical dimensions of the text. At the narrative level, Sarrasine's catastrophe registers as a formal separation between the two parts of the whole: the first is situated at the Lanty's, where the narrator and Madame de Rochefide observe mysteries in need of resolution; the second is situated at Madame de Rochefide's, with the story about Sarrasine's childhood offering a means of solving the mysteries posed in the first. The narrative "catastrophe" corresponds exactly to the historical rupture between Old Regime Catholicism and the post-revolutionary period, effectively between two countries, France and Italy, and between two cities, Paris – the capital of modernity – and Rome, "la patrie des arts", but also the heart of Catholicism. The abrupt crossing of temporal and geographical frontiers corresponds to the internal rupture of consciousness marking the separation between Sarrasine and the narrator. Attached to Catholic France (figured geographically by the still Catholic nineteenth-century Italy where Sarrasine discovers transcendent or "divine" love), but situated in post-revolutionary, post-Catholic France (figured by Paris, "la société sans croyance"), the narrator's fragmented "je" is not simply an "other" (to invoke the famous line of Rimbaud "je est un autre"), it is rather an *other country*.

The peculiar position of the narrator illustrates the point: situated in the window-frame of the Lanty mansion, divided between two worlds, two perspectives and two feelings, his dual perspective reflects an internal duality of consciousness:

> Ainsi à ma droite, la sombre et silencieuse image de la mort; à ma gauche, les décentes bacchanales de la vie [...]. Moi, sur la frontière de ces deux tableaux si disparates, qui, mille fois répétés de diverses manières rendent Paris la ville la plus amusante du monde et la plus philosophique, je faisais une macédoine morale, moitié plaisante, moitié funèbre. (pp. 227-8)

The narrator does not immediately reveal the content of these two disparate worlds which he views as a banality of modern Parisian life, nor does he explain the psychological division between life and death or between the exterior and interior worlds. This internal division may be connected to his catastrophic experience with the old man since his perspective on him is also radically divided. On the one hand, the old man is simply "le vieillard" (which the narrator reiterates almost as if to reassure himself of this fact). On the other, the old man is dead, in ruins, a spirit, a ghost, a source of cold, darkly-clothed and smelling of a cemetery. What the narrator sees with his eyes (an old man) corresponds neither to the emotions nor to the images that the narrator associates with the old man's younger incarnation that he once knew (la Zambinella). An understanding of the historico-cultural, and even "religious" stakes attached to the narrator's crisis of consciousness will require a thematic and symbolic interpretation of the old man.

At the thematic level the narrator's divided psyche has an identifiable cause: the disillusionment and symbolic death that he underwent as Sarrasine. The anonymous narrator, a double of Sarrasine, emerges resuscitated from death and endeavours to explain the cause and the consequences of the spiritual catastrophe to others. His narratorial dilemma is that his nineteenth-century reading-public will be perplexed or scandalized by his love object, since he had fallen in love with *a man*. Moreover, the love that la Zambinella revealed to Sarrasine was believed to be transcendent and even "religious". It is unclear how a modern and secular reader grounded in rationalism and/or materialism would understand Sarrasine's divinized love. The narrator is hyperconscious of the epistemic rupture between his mystified consciousness as "Sarrasine" (which he is attempting to reconstruct for us) and the rationalistic understanding of his modern interlocutors.

Understanding this lost illusion involves great difficulty, and is not without certain dangers. In order to avoid an immediate scandal and gain the confidence of modern readers, Sarrasine takes cover under his own death, doubling and obscuring himself behind the anonymous narrator while hiding the identity of his ideal love object behind a feminine appearance. In other words, he transforms his loss of religious love into a hoax love story, recounted anonymously and in the third-person, about how Sarrasine fell in love with an opera singer, a castrato disguised as a woman. At the thematic level, we could easily conclude that the narrator does not master his story. Certain critics have argued that since the narrator scandalized Madame de Rochefide to the point of refusing to "pay up" for the story (they were to be lovers if the narrator revealed the mysteries of Part I), the narration is a failure. Barthes, for example, considers the narrator's failure to conclude this deal as evidence of the triumph of "l'infini du langage" over mimetic realism. But is the scandal awaiting Rochefide (i.e. the revelation that la Zambinella is a man, that the old man is an aged version of la Zambinella or that the painting of the Adonis is la Zambinella in his youth) identical to the scandal that Balzac prepares for his reader? For what narrator would publicly recount his own failure if there were not some hidden and more serious objective? In our view, the narrator's failure plays an important role in Balzac's narrative strategy, serving as a vehicle for a deeper novelistic communication. Seen from this angle, the "death" of Sarrasine (the young narrator) dramatically marks the temporal and epistemic frontier between his mystified (and "crazy") state of mind and his consciousness after the disillusionment.

That the obvious scandal (i.e. that la Zambinella is a man) communicated by an openly ironic narrator hides another, more serious, scandal (the hidden identity between the narrator and Sarrasine and the secret declaration of his homoerotic love for la Zambinella) would explain why the narrator so carefully reviews the elements of his (lost) illusion when he sees the old man, after many years of separation, at the Lanty mansion. The old, displaced emotions associated with la Zambinella surge forth into consciousness, reminding him of both the illusion of ideal love and its catastrophic loss. The entire story of "Sarrasine" is then a therapeutic retelling of these repressed memories.

Evidence that the narrator's consciousness is a double of Sarrasine's is apparent in the laborious description of the old man, which seems to lack significance, yet which, on examination, reveals the narrator's emotional attraction or disturbance. Is a certain nostalgia for the passion that he had known before his disillusionment not evident in the simple fact that the narrator is "attristé de voir tant de ruines [...]"(p. 235)? Why would this be if he had not been formerly so attached to the totality? And who else in his tale other than Sarrasine had known la Zambinella in his perfection and would thus be in a position to regret his current ruined state?

There are other examples of a residual attachment: who other than Sarrasine would see in the old man the image of a woman (we know that Sarrasine had first perceived la Zambinella dressed as a woman)? The narrator reveals that he had once closely scrutinized the young version of the old man's body: "Son excessive maigreur, la délicatesse de ses membres, prouvaient que ses proportions étaient toujours *restées* sveltes" (p. 234). Without appealing to the myth of the omniscient narrator, who except Sarrasine would have known this body sufficiently to make a comparison between present and past? And who other than Sarrasine would associate la Zambinella symbolically with a Christian or dead Christianity: "Vous eussiez dit deux os mis en croix sur une tombe" (p. 234). The Christological allegory of love will be discussed later.

Let us now reflect on evidence of an identity between Sarrasine and the narrator, or, more precisely, of Sarrasine as the "other-of-the-self" of the narrator. Consider, for example, the strange event that takes place immediately after the description of the old man: the narrator personifies his thought and underscores this act so that the reader does not fail to reflect on its content. The personified thought is that of a union between an old man and somebody who is twenty-two years old, and between life and death: "Par un des plus rares caprices de la nature, la pensée de demi-deuil qui se roulait dans ma cervelle en était sortie, elle se trouvait devant moi, personnifiée, vivante, elle avait jailli comme Minerve de la tête de Jupiter… elle avait à la fois cent ans et vingt-deux ans, elle était vivante et morte" (p. 233).

The "union" and the identity of the two people are only hinted at by the narrator. The two most obvious referents are the old man and Marianina, described by the narrator as follows: "Voir, auprès de ces débris humains, une jeune femme...: ah! c'était bien la mort et la vie, ma pensée, une arabesque imaginaire, une chimère" (p. 236). In a crucial scene, the old man offers a ring to Marianina, to which Madame de Rochefide reacts: "Est-ce son mari? Je crois rêver. Où suis-je?" (p. 238)

But to conclude that the twenty-two-year old half of the union is Marianina does not resolve anything since the symbolic marriage between the old man and Marianina, occurring just a few pages later, is itself a piece of the puzzle. The most important question asked by Madame de Rochefide ("Est-ce son mari?") about the ring exchange is precisely the catalyst that prompts the narrator to recount Sarrasine's past as a key to solving the mystery of the marriage. The enormous irony of "Sarrasine", however, is that narrator never manages to solve this mystery.

By this strange omission Balzac may be simulating a blockage in order to make the reader look for the solution, requiring us to follow the allegorical logic of the tale. Described as "le type de cette poésie secrète, lien commun de tous les arts, et qui fuit toujours ceux qui la cherchent" (p. 229), Marianina plays a fundamental but hidden role, according to Balzac, in his own artistic work. The narrator wants us to discover his union with the old man, but cannot refer to himself or to the union directly without causing a scandal. Marianina, personifying a resuscitated Sarrasine, provides the means to present to the public a heterosexual union, but by linking it to Sarrasine the anonymity of the narrator remains intact. If the narrator manages to make us believe that the two sides of the union of his personified thought are the old man and Marianina, it is through a logic of contiguity, but also because bourgeois conventions push us to find a union between man and woman as a solution to the marital mystery. Marianina, as the only young woman in the story, is therefore assumed to be the twenty-two-year old symbolic spouse of the old man.

The kernel of the entire narrative, the union on which all the other events depend, but which remains nonetheless obscure, is the union that Sarrasine fantasizes about with la Zambinella at the opera in Rome, of which the memory is unsignifiable in a word. Balzac puts us onto the scent by various indirect means, including literary tropes and figures (allegory, personification, metaphors) and by use of an archaeologically structured narrative: the reader's task is to piece together the fragments of the narrator's lost illusion from an another era.

Returning to the question of the twenty-two-year-old half of the personified thought, Sarrasine is himself precisely this age when he discovers la Zambinella's love in Rome. Is this pure coincidence, or an important piece of evidence in the secret story of the union between the narrator / Sarrasine and la Zambinella? For it is the narrator who insists on this detail, and why would he be so precise in stating the age of the partner in both unions with la Zambinella if no secret link existed between the two? Marianina, in our view, figures Sarrasine in the modern, bourgeois era. The narrator needs her feminine identity in order to indicate discreetly to the modern and bourgeois reading public the scandalous union of Sarrasine and la Zambinella, while giving us the key to deciphering his identity in her. When the narrator sees the old man for the first time after so many years, emotions and memories attached to the young Zambinella rise to consciousness alongside his current appearance, leaving the narrator divided between two images and two sets of emotions. La Zambinella's reality is contained in neither of the images: it is, strictly speaking, ruptured by the historical dichotomy. Yet no scientific, historiographic or narrative convention existed at the time to capture this split reality of his consciousness, especially with one half anchored in an admissible love. It falls to the reader to archaeologically reconstruct the past from the present fragments of the narrator's fractured psyche.

Consider now the key scene where Sarrasine imagines himself in a mystical union with la Zambinella. Rochefide's question ("Est-ce son mari?") is reformulated, reminding us of the narrator's original objective in telling the story of Sarrasine, i.e. to solve the mystery. We begin to perceive that la Zambinella symbolizes a spiritual or religious kind of love for Sarrasine.

First, the context: the love is "revealed" to Sarrasine while in Rome, and more precisely, in the Papal territory. Balzac wrote "Sarrasine" soon after the Revolution of 1830, that is, just after "l'effondrement catastrophique" of the Restoration, in a France separated from its Old Regime / Catholic foundations – a rupture originating in the eighteenth century but marked dramatically by the Revolutions of 1789 and 1830. Situating Sarrasine's religious experience of mystical love at the heart of Catholicism in Papal Rome, Balzac figures more clearly the problem of "les deux Frances" in the nineteenth century, allowing readers to imagine a religious content in Sarrasine's love for la Zambinella. The voyage in geographical space is then simultaneously a voyage in time, towards a Catholic homeland that remains intact. The "voix céleste" of la Zambinella attacks Sarrasine's soul (p. 244), penetrates his pores (p. 242) and is received as "[une] révélation". The erotic symbolism of this penetration is inspired by the Christian/mystical tradition according to which spiritual operations (in this case, the spiritual union between the souls of la Zambinella and Sarrasine) are figured by erotic metaphors such as sexual union, orgasmic ecstasy, etc. On the one hand, we find a chain of religious themes and metaphors characterizing Sarrasine's experience (la Zambinella sings for the Catholic church, his friends are Christians, he possesses "[une] grâce inimitable", and "[une] grâce infinie" (p. 243). On the other, we find a mystical eroticism: the penetration of la Zambinella's voice in the soul of Sarrasine produces an orgasmic ecstasy that provokes "des cris de plaisir" (p. 243) and leads him to imagine "des principes nouveaux de l'existence" (p. 244).

This scene, which is the key to all of Rochefide's questions (and thus to understanding the fragmented consciousness of the narrator) remains a total mystery if we miss the connection between the marriage between the old man / Marianina and that of Sarrasine / Zambinella. To underscore its importance, the narrator returns to Madame de Rochefide's question about the old man / Marianina's marriage right after he describes the experience of mystical union between Sarrasine and la Zambinella at the opera: "Mais, me dit madame de Rochefide en m'interrompant, je ne vois ni Marianina ni son petit vieillard". The narrator replies: "Vous ne voyez *que lui*! m'écriai-je, impatienté comme un auteur auquel on fait manquer l'effet de coup de théâtre" (p. 245, our emphasis).

This is the same theatrical mechanism featured earlier in the reference to the thought of a united couple resurrected in the narrator's consciousness. The memory of this mystical union in Rome surges into the narrator's consciousness in Paris, though he does not tell us this directly, making his reader discover the connection via indirect clues. Rochefide's confusion and frustration constitutes Balzac's admission that his readers are undoubtedly experiencing similar confusion as we move through the narration in search of meaning. The narrator's explanation to Rochefide that she sees "only him" is also for the reader. Most likely, the narrator imagines la Zambinella as half of a mystical union. According to the conventions of Catholic mysticism, the union of two souls and two fleshes forms a perfect unity. Translated visually, it would be perfectly logical to see only one body. Hence, the narrator's exclamation: "Vous ne voyez que lui!"

Analysis of the other piece of mental debris, the painting of the Adonis, confirms this reading. The painting would probably pass unnoticed as a puzzle piece for reconstructing the narrator's consciousness without Madame de Rochefide's passion for the body that is figured in the painting and without her implausible question: "Qui est-ce?" The narrator knows the identity of the model and promises to reveal it to her. The text indicates that the painting is a copy by Girodet ("Endymion") of another work by Vien, which, in turn, is a copy of the statue sculpted by Sarrasine. Balzac thus uses a "real" painting in order to give the reader external access to the mental state of Sarrasine while under the illusion of idealized love.

We see a hermaphrodite with both male and female traits, but whose gender is, ultimately, masculine. Does this mean that the original model for all the copies is that of a man? The narrator explicitly says that the statue is of a woman while contradicting himself with the presentation of the material evidence and narrative symbolism, thus casting doubt on his own reliability.

Yet if the narrator is Sarrasine, he cannot expose his love story openly. For epistemological and moral reasons, he is forced to communicate indirectly, playing a double game by presenting his union in heterosexual terms while offering clues about the concealed truth. The statue is not a purely visual entity, but emerges from the union of a phallic voice (the symbol of idealized masculine love) penetrating the vaginal soul of Sarrasine. The sculptor exteriorizes (personifies) this interior "spiritual" union by figuring it as a human body. And why not a male, since the erotico-mystical images representing the voice of la Zambinella are phallic while the soul of Sarrasine is female?

Access to Sarrasine's archaic state of mind may also be found in Balzac's metaphorical descriptions of Sarrasine sculpting. The narrator seems to indicate Sarrasine's unconscious awareness that the ideal woman is phallic, and his success in capturing this masculine dimension of la Zambinella in his statue. Though this initially seems implausible, Sarrasine's mental image of la Zambinella emerges not only from his visual observation but also from what he hears, as the following citations indicate: "Pendant une huitaine de jours, il vécut toute une vie occupé le matin à pétrir la glaise à l'aide de laquelle *il réussissait à copier la Zambinella, malgré les voiles, les jupes, les corsets* […] qui la lui dérobaient" (p. 245, our emphasis). "[I]l se familiarisa graduellement avec les émotions trop vives que lui donnait le chant de sa maîtresse; puis il apprivoisa ses yeux à la voir" (p. 245). What Sarrasine "sees" in the body of la Zambinella is filtered by his imagination and by *his ears*; the emotion generated by la Zambinella's voice permits Sarrasine to imagine his body in all its truth.

That Sarrasine unconsciously desired a man's love should not shock since any Christological allegory figures divine love, the love that gives life, as a love communicated by a male – God, Christ. In the modern and desacralized context in which the narrator tells the story of his lost illusion, mystical love no longer officially exists and is, in any case, incomprehensible according to a rationalist or materialist epistemology. This explains the scandal of signification generated by the relationship between Sarrasine and la Zambinella. At one level, a love between two men registers in a modern context as homoerotic love. At another, any attempt to signify a religious or transcendental ideal will necessarily cause a slippage of meaning and infinite signification: immanent linguistic conventions cannot, by definition, render divinity. It registers as an absence or gap in meaning.

Thematically, Sarrasine's catastrophe registers as a purely psychological event that took place, apparently at the moment when the sculptor understands that his ideal love is a hoax and that la Zambinella had only "played" at love. It is also the moment at which Sarrasine understands that persisting in the hope of realizing ideal love would be "une folie" (p. 257), whereas renouncing it would kill him. The solution is to break his consciousness in two where one part must "die" and the other survives to recount the cause of the death.

But why would la Zambinella's love, or rather its absence, trigger Sarrasine's death? Sarrasine had interpreted la Zambinella's love as one that could give life, by which he means transcendent, eternal life. This is why Sarrasine exclaims when he discovers that he is not loved: "Monstre! Toi qui ne peux donner la vie à rien [...]" (256). From this perspective, we should expect a spiritual death at the moment he is disillusioned by la Zambinella's lack of a heart. Sarrasine "expires" when he is stabbed by la Zambinella's Christian friends, but why would Christians kill Sarrasine, especially when he was already the victim of their hoax? In our view, Sarrasine's death is symbolic, marking a psychological or spiritual catastrophe in which his psyche is split between life connected to spiritualized, Catholic love and a disenchanted, materialist life without it.

Most readers think that Sarrasine's disillusionment comes from the fact that la Zambinella is not, after all, a woman: he thought he loved a woman, but it was a man. The narrator, however, plays a double game, revealing that Sarrasine knew his ideal object of love to be a man. What disillusioned him – on this point the text is clear – was the absence of ideal love: "[E]n fouillant ton être avec cette lame, y trouverais-je un sentiment à éteindre, une vengeance à satisfaire? Tu n'es rien. Homme ou femme, je te tuerai! [...] Plus d'amour! Je suis mort à tout plaisir, à toutes les émotions humaines" (p. 256).

The absence of love provokes a catastrophic separation, a death in the soul, a radical disillusion, after which the narrator will no longer have direct access to the soul or to the transcendent love of la Zambinella, or to his own former state of mind. While remaining sentimentally attached to his former self (the self that had access to divine love), his rational consciousness and his language become radically detached. Sarrasine's madness and death thus indicate not only an epistemic separation between Sarrasine and Zambinella, but also between the young Sarrasine and the older Sarrasine-narrator.

Up to this point our allegorical reading remains speculative. However, we find confirmation at the end of the narrative, when Madame de Rochefide interprets for us what the narrator has been attempting to accomplish, sharing his religious disillusionment with her in order to shake her from her Christian illusions:

> Ah! s'écria-t-elle [...]: Vous m'avez dégoûté de la vie et des passions pour longtemps. Au monstre près, tous les sentiments humains ne se dénouent-ils pas ainsi, par d'atroces déceptions?… Si l'avenir du chrétien est encore une illusion, au moins elle ne se détruit qu'après la mort. (p. 257)

Balzac exposes here the historical and religious content of Sarrasine's illusion. According to Rochefide, the narrator/Sarrasine had erred in yielding to disillusionment whereas she will persist in her belief until death, even if, from a rationalist point of view, it is madness: "Paris, dit-elle, est une terre bien hospitalière; il accueillit tout… Le crime et l'infamie y ont droit d'asile; la vertu seule y est sans autels. Oui, les âmes pures ont une patrie dans le ciel! Personne ne m'aura connue!" (p. 258).

A reading of "Sarrasine" as an archaeology of consciousness provides a better way of understanding why la Zambinella remains unnameable and without substance for the narrator. The divine love and religious belief that la Zambinella symbolizes have no real substance in a modern, post-Catholic world: this is why the old man appears to the narrator as a ghost with a dead and hollow body. Since Sarrasine's experience of this transcendent reality is no longer immediately accessible to his disenchanted consciousness, the narrator's problem is to make this experience intelligible to a reader whose modern consciousness depends precisely on the repression and perpetual displacement of the object – religion – that Balzac is attempting to expose. The narrative solution to this experience of catastrophic rupture is none other than to apply to consciousness an archaeological method whereby: "La cause fait deviner un effet, comme chaque effet permet de remonter à une cause" (X, 657).

Archaeology – A Passion of George Sand

Claire Le Guillou

Abstract: *In contrast to Balzac, George Sand's engagement with archaeology was a very practical matter. Throughout the 1850s she collected artefacts retrieved from archeological digs, financed excavations in her local Berry and took part in other archaeological ventures in the area. This interest necessarily translated into her later works, in the characters of archaeologists and antiquarians that figure in them and also, the chapter argues, in the construction of narrative and in the inscribing and "unearthing" of the values of a much earlier Stone and Iron Age France in her fiction of the period.*

George Sand's intellectual curiosity was quite dazzling. She threw herself into the study of botany, geology, entomology and, although it is less widely known, archaeology and one of its branches in particular, numismatics. Her archaeological interests have often been overlooked by critics,[1] due to the fact that they have been either largely hidden by her research into folklore and ethnography or have, alternatively, been understood as forming part of those activities.[2]

---

[1] See, however, Marie-Louise Vincent, *Le Berry dans l'œuvre de George Sand* (Marseille: Laffitte Reprints, 1978), Chapter II. Equally worthy of note are the two following articles, written by the archaeologist Gérard Coulon, to whom our grateful thanks are due: "George Sand et la passion de l'archéologie", *Magazine littéraire*, janvier (1992), n° 295, p. 47, and "George Sand et l'archéologie", *La Lettre d'Ars*, mai 1996, n° 2, pp. 5-6.

[2] A. van Gennep, "George Sand folkloriste", *Mercure de France*, 1er juin 1926, no. 671, and Daniel Bernard, "George Sand pionnière de l'ethnographie", in *George Sand, une Européenne en Berry* (Le Blanc, Châteauroux, la Châtre: Amis de la Bibliothèque municipale du Blanc et Comité du Bicentenaire George Sand, 2004), pp. 121-49.

George Sand and archaeological relics

Her interest in and her passion for archaeology was apparent in her many site visits and, most especially, in her excavations. It should, however, be noted that she was particularly partial to Celtic archaeology. While she took the opportunity to visit Roman sites in the south of France and in Italy, these visits simply did not hold the same fascination for her.[3] George Sand needed a more immediate understanding of places, regarding monuments from the past as open books permitting a privileged reading of history. In short, archaeology's true attraction for her lay in its capacity to enrich her appreciation of specific locations and to reinforce the ties that bound her to everyday life in her own small universe in the regions around Berry and la Marche. As Nicole Belmont notes, Sand's rejection of Greco-Latin civilization is merely a natural consequence of her passionate preference for Celtic antiquities.[4] In that respect, George Sand very much followed in the tradition of the Académie Celtique.

By the 1830s, druid stones were already proving to be a source of interest for the novelist. In 1837, in *Mauprat*, a very short scene takes place on a druid stone at Crevant. This site would become a favourite beauty-spot for the whole Sand family in the 1850s, where they would go to pick flowers and to catch butterflies among the standing stones. Subsequently, in 1841, Sand visited a much more impressive site, that of Toulx-Saint-Croix, in the Creuse region. There she discovered the standing stones, taking great pleasure in their contemplation. She wrote to Eugénie Duvernet that this journey "[lui] a donné l'envie de faire un roman sur Toul et le maître d'école". Then, she continued her letter in the following terms: "J'aurai besoin de notions géographiques et statistiques plus exactes que celles que *je n'ai pas prises*. Je compte sur Charles pour cela."[5] From this short visit and from her reading and information gathering, there would emerge three years later *Jeanne*.

---

[3] *Correspondance de George Sand*, ed. by George Lubin, 29 vols (Paris: Garnier, 1976-1979), XIII, pp. 116-17.

[4] Nicole Belmont, "L'Académie Celtique et George Sand, les débuts des recherches folkloriques en France", *Romantisme*, (1975), n° 9, pp. 29-38.

[5] Sand, *Correspondance*, V, pp. 472-73.

Then, in the 1850s, a stroke of good fortune would have it that archaeological excavations were undertaken on her very doorstep. This opportunity led, during the month of February 1853, to George Sand becoming a real archaeologist herself. Following the discovery of burial caskets and human bones close to the church at Vic, Sand financed the dig herself, then undertook others in the grounds of the château at Nohant. Things got under way on 10 February, as witnessed in the author's notebooks: "Sol a été aujourd'hui à Vic où l'on a trouvé des cercueils de pierre et des ossements; elle a rapporté une mâchoire pour faire la vaillante. Ce soir, elle en a eu une peur de chien et a voulu l'ôter de sa chambre, mais elle ne l'a pas retrouvée et a eu d'autant plus le trac."[6] From then on, the notebooks and correspondence of George Sand constitute a veritable excavation diary in which each stage of her archaeological research activities is recorded in minute detail.

Several days later, on 14 February, it would be the turn of Emile Aucante to visit Vic: "Emile a été à Vic chercher un cercueil de pierre; il a rapporté une pierre tumulaire et un autre petit cercueil d'enfant, un autre aussi cassé, puis des perles en verroterie de couleur et une clef, le tout provenant des fouilles de Vic. Mme se propose d'en faire faire à ses frais."[7] The following day, there was a new expedition to discuss the formalities relating to the excavations to be undertaken: "Manceau et Emile vont à Vic convenir avec le curé et Mr Aulard des endroits où Mme pourra faire des fouilles à ses frais. Le soir, on lit force livres d'histoire pour tâcher de découvrir l'origine de quelques monnaies prêtées par le curé et trouvées dans les fouilles de l'église."[8]

On 16 February, the author, having taken charge of the dig, finally appeared on the site, accompanied by her grand-daughter:

> Après déjeuner, elle va à Vic avec Nini, Emile et Manceau pour voir les fouilles qu'elle fait exécuter à ses frais autour de l'église. Comme il y fait froid, elle revient avec sa petite fille, après s'être réchauffé chez les bessons. Manceau et Emile restent, aidés de Jean Brunet, Jacques Soulat et Pajot. Au niveau du sol de l'église (en dehors) on trouve beaucoup de cercueils de pierre qui ont été ouverts et vidés. Audessous, le terrain est vierge de toute violation et l'on trouve 3 et 4 couches de cercueils superposés. Ceux qui sont immédiatement au-dessous de ceux de pierre ne font qu'entourer la tête, le corps est libre et à même la terre; la tête est recouverte

---

[6] George Sand, *Agendas*, 5 vols (Paris: Touzot, 1990-1993), I, p. 88.
[7] Sand, *Agendas*, I, p. 89.
[8] Sand, *Agendas*, I, p. 89.

d'une pierre très lourde. Au-dessous de ces derniers, on trouve les cercueils en maçonnerie, recouverts de pierres grossièrement taillées; des squelettes sont parfaitement conservés.[9]

This same day, Sand recounted her archaeological activities to her son:

Je vais bien aujourd'hui et j'ai été me promener jusqu'à Vic, où l'on retourne le terrain autour de l'église et où l'on trouve des tombeaux et des ossements comme si toutes les armées de César et autres Ostrogoths y avaient passé. J'ai fait apporter trois cercueils de pierre dans notre jardin et avec la permission du maire et du curé j'ai mis trois ouvriers pour remuer un petit coin, où l'on a trouvé aujourd'hui que des débris déjà fouillés à je ne sais quelle époque. En fouillant plus bas, au dessous de la couche des sarcophages, on trouve de la brique romaine, et des squelettes couchés avec ordre dans des cercueils de maçonnerie, la tête couverte seulement d'une pierre. Mais pour faire faire des fouilles avec soin l'endroit n'est pas commode et nous n'avons trouvé ni monnaie ni bijoux. Mais ces découvertes nous ont mis en goût de recherches, et comme je me rappelle un endroit du jardin sous les noyers, d'où j'ai vu extraire autrefois toute une première couche de sépultures et d'ossements, nous allons nous amuser à faire creuser plus bas pour voir si là aussi nous trouverons le lit romain. Alors, en y ayant l'œil et la main, nous trouverons peut-être des monnaies et des lacrymatoires.
Pendant que nous fouillons les tombes et qu'Emile [Aucante] penché sur la *fosse béante*, se donne des airs de vampire, tu cours le bal et la mascarade.[10]

The following day, excavations were also undertaken at Nohant: "Le jardinier n'ose plus entrer seul, le soir, dans sa serre parce qu'il y a des sarcophages à la porte. On a commencé à fouiller autour des noyers du jardin; on trouve une masse de squelettes, mais à même la terre, pas la moindre trace de maçonnerie. Emile a rapporté ce matin des fouilles de Vic une petite médaille romaine."[11] Snow and inclement weather interrupted the archaeological dig for several days, during which Sand took the opportunity to tell her daughter about the excavations:

Je te dirai que l'ardeur des fouilles s'est tellement emparée d'Emile et de Manceau que la maison est pleine de cercueils de pierre et d'ossements, si bien que le jardinier a une peur de chien et n'ose plus aller fermer sa serre à 7 h. du soir, sans être accompagné de Caillaud, et encore foire-t-il dans ses chausses. J'imagine que tu en ferais autant. Mais voilà bien autre chose. Pour satisfaire la passion de ces jeunes

---

[9] Sand, *Agendas*, I, pp. 89-90.
[10] Sand, *Correspondance*, XI, pp.592-93.
[11] Sand, *Agendas*, I, p. 90.

antiquaires, j'ai indiqué un endroit sous les noyers du jardin que je sais être l'ancien cimetière du village et je leur ai fait cadeau de quelques journées d'ouvriers pour en avoir le cœur net. Depuis ce matin, on a trouvé déjà un lit de squelettes avec des débris de petites urnes. Aussi la situation devient-elle des plus fantastiques et Manceau assure-t-il que le *sol sur lequel l'homme marche n'est qu'un vaste ossuaire*. Emile commence à avoir l'esprit frappé tout en mangeant de la perdrix à l'estouffade, il prétend que le cœur lui tourne et qu'il croit manger des ossements humains, d'où il conclut que *le beau c'est le laid* et que *la foi, c'est l'athéisme*. Il dit là-dessus de si belles choses à dîner, accompagnées de rires mystérieux et sataniques, que je commence à craindre l'effet du pavillon *fatal*, sur sa cervelle.
Nous avons pour toi des perles et une épingle. C'est fort laid mais cela sort des os pourris et c'est fort romantique. Nous n'avons encore découvert dans tout cela qu'un fait curieux. C'est qu'on a beau creuser, on n'arrive pas à trouver la dernière couche des sépultures. Elles sont par lits, les une sur les autres, sans fin.
Mais voilà assez d'ossements, et je te vois déjà le cauchemar. Nini s'en moque et regarde tout cela avec la plus grande indifférence. Les habitants disent que nous cherchons des trésors, et un liard romain que nous avons trouvé sur la tête d'une dame leur a paru mériter beaucoup de commentaires.[12]

Her notebooks bear witness to the recommencement of the dig on 24 February, and indicate several interesting finds: "Les fouilleurs trouvent une espèce de bague, 2 urnes lacrymatoires, 4 monnaies du temps du roi Jean (nous croyons), une des monnaies avait, en s'oxydant, conservé intacte sur son effigie et son revers la portion du suaire qui l'enveloppait."[13] The following day, the dig yielded "2 monnaies, dont une de 1594, plusieurs urnes". The day after, excavations proved fruitful once again. They found "des urnes, 3 bagues de fonte, 2 pièces de monnaies".[14] The day's discoveries were once again recorded on 28 February. These included seven coins, some broken urns and three stone burial caskets. The major interest of this event is, however, the accompanying drawing, which shows the cross-section of a tomb with a skeleton laid out on a bed of clay and covered with soil and stones. The crowning moment of these excavations took place on 1 March, thanks to the discovery of "une urne rouge avec des ornemens [*sic*]".[15] The next day, Sand wrote once again to her son to inform him about the progress of the dig and to offer him several drawings of the excavations, accompanied by one of her articles for *L'Illustration*. Unfortunately, the project was never completed, much

[12] Sand, *Correspondance*, XI, pp. 594-95.
[13] Sand, *Agendas*, I, p. 92.
[14] Sand, *Agendas*, I, p. 92.
[15] Sand, *Agendas*, I, p. 93.

like the excavations which were to finish definitively on 3 March with "rien de nouveau".[16] These initial excavations, which took place over more than three weeks and which were the object of daily discussions, bear the marks both of romanticism and of the fantastical. Emile Aucante, leaning over a "fosse béante" becomes the romantic figure of Hamlet, as depicted by Eugène Delacroix. But what George Sand emphasizes particularly is the fear that the archaeological finds aroused in the peasants and the ordinary folk, as well as the air of madness that they brought with them to Nohant. Sand thus experienced directly the fantastical dimension of archaeology.

Her next dig would take place under less extravagant circumstances. In 1857, Sand undertook a new archaeological expedition at the cairn at Presles, in order to find inspiration for the setting of her historical novel *Les Beaux Messieurs de Bois-Doré*. From 4 January onward, her notebooks are filled with reflections on the origin and the nature of la Motte de Presles, which she shared with her son on the 24 of the same month:

> Que sont les vestiges incompréhensibles auprès du *tumulus*? La motte de Presle, c'est-à-dire un petit fort; mais que défendait-il? Il est trop loin du château du Magnet, pour l'avoir jamais protégé. Il défendait donc la ville? Car il y avait une ville, toutes les histoires du Berry en font mention. Où était-elle? Comment et quand a-t-elle disparu? Il y a eu, dans ce petit fort, une bataille sérieuse. Ces pièces noircies par le feu et ces ossements qu'on trouve en si grande quantité dans le sable, prouvent meurtre et incendie. Les débris de poterie mêlés à ces ossements, je les explique ainsi: au 16e siècle on a employé la *pluie des boulets d'artifice*, faits avec des écuelles de bois et des *pots de grès* qui incendiaient en éclatant.
>
> Tout cela ne me dit pas quand et comment ce fort et cette ville ont disparu. Il en est encore question au 16e siècle, et ensuite *néant*.[17]

On 15 January, ostensibly with a view to confirming her hypotheses, Sand embarked upon the project of excavating la Motte de Presles, a project which was still on-going on 17 February, as we learn from Alexandre Manceau in the diaries: "Mme va à la Motte de Presle avec Manceau, Emile, Sylvain et Gabriel [Blain]. Gabriel pioche comme un enragé et nous trouvons des morceaux de poteries, des tuiles, des briques et des os (un cadavre entier) et puis encore des morceaux de poterie, des tuiles, des briques et des os."[18] Thereafter,

[16] Sand, *Agendas*, I, p. 93.
[17] Sand, *Correspondance*, XIV, p. 192.
[18] Sand, *Agendas*, II, p. 11.

archaeology became an element of the author's work, for which her passion did not, over the course of the coming years, wane. In the 1860s, her interest in numismatics was, for instance, still apparent. Jean-Claude Rouet, civil engineer at La Châtre, had a gold Roman coin sent to Sand in January 1863. She thanked him in these terms: "Mon cher Rouet, votre monnaie est très belle, très [bien] conservée. Je ne sais pas encore si elle est rare, je le crois, mais je n'ai pu l'étudier aujourd'hui. Dans tous les cas, ces trouvailles ont un très grand intérêt et je vous remercie de cet aimable cadeau qui m'est très agréable. Vous me direz où elle a été trouvée, car j'espère bien vous voir dimanche soir et vous serrer la main."[19] In this extract from her letter, the author reveals the instincts of a meticulous practitioner of archaeology, her first reaction being to ask where the coin had been discovered. Then, with the intention of identifying it, she wrote on the 21st of the same month to Louis Maillard: "Informez-vous d'un ouvrage de numismatique à m'envoyer pour les monnaies et les médailles celtiques, gallo-romaines et romaines, le meilleur possible avec planches et pas trop cher."[20] The following week, she repeated her request for the said work.

Sand's archaeological interests extended into the 1870s, and led to her finally discovering the "mardelle" that she had sought so avidly. "Mardelles" are excavations in the ground, which are widely considered nowadays to be the result of karstic phenomena (underground water carves out channels and caves susceptible to collapse from the surface; eventually a sinkhole, or a depression formed by a portion of the lithosphere below eroding away, is created). However, in George Sand's day, mardelles posed other serious archaeological problems. In his *Ethnogénie gauloise*, for instance, Roger de Belloguet considers them as indications of the underground sections of ancient Gaulish dwellings. Consequently, the Sand family were most enthusiastic in seeking out these landforms, guided by toponyms such as "La Mardelle", "Les Mardelles", "Le Bois des Mardelles", etc. Near to Maron there is just such a plot of land, named "La Mardelle",[21] which was the object of a family expedition on 18 September 1862: "Maurice, Lina et Cadol vont faire une g[ran]de prome-

[19] Sand, *Correspondance*, XVII, p. 369.
[20] Sand, *Correspondance*, XVII, p. 391.
[21] See Stéphane Gendron, *Les Noms de lieux de l'Indre* (Châteauroux : Académie du Centre et Credi éditions, 2004), p. 51.

nade à la recherche des *mardelles*. Ils ne trouvent qu'un tumulus à Marron."[22] The prized "mardelle" was not finally discovered until 1873, as Sand informed her daughter on 19 April of that year: "Nous avons enfin découvert une *mardelle* dans la brande. Il y a cinquante ans que j'en cherche et la brande en est remplie. Mais il faut être dessus pour les voir."[23] While it is tempting to conclude that unskilled prospecting techniques may have been responsible for this lengthy pursuit, it has been independently noted that: "La recherche de ces excavations est bien plus difficile qu'on ne pourrait le supposer au premier abord. On ne les aperçoit pas de loin, elles sont souvent même dissimulées dans les replis du terrain, au milieu des fourrés, dans les bois."[24] Otherwise, throughout her life, Sand frequented a number of antiquarians and archaeologists, and the catalogue of her library bears witness to her abiding interest in their works, among which the writings of Pictet, Feydeau, de Belloguet and Raynal feature prominently.

Archaeological inspirations in Sand's œuvre

Sand's obvious enthusiasm for the study of antiquities poses the question of how archaeological themes actually influenced her writing. Can we, for instance, discern the presence of standing stones, ancient currencies or *mardelles* in her work? What use does she make of archaeology? How does she present the process of archaeological discovery in her writings? And how does she describe the figure of the archaeologist for her readers? Firstly, although the traces of archaeological influence in her works are, at times, extremely subtle, they reappear constantly throughout the duration of her career, thus bearing witness to a long-standing preoccupation on her part. Several titles suggest themselves immediately: *Jeanne* in 1844, *Le Péché de Monsieur Antoine* in 1847, *La Famille de Germandre* in 1861 and *Nanon* in 1872. Nor should we forget the short story "Le Marteau rouge",

---

[22] Sand, *Agendas*, III, p. 59.

[23] Sand, *Correspondance*, XXIII, p. 142.

[24] *Congrès archéologiques de France*, XLe session, séances générales tenues à Châteauroux en 1873 (Paris: Derache-Didron-Dumoulin Libraires, 1874), p. 150. See also Geneviève Dindinaud, "Sur les 'mardelles' de Dun-sur-Auron", *Cahiers d'archéologie et d'histoire du Berry* (1969), 22-26.

published in 1875, of which a necessarily brief overview will be given.

By way of preliminary considerations, it should be noted that these few novels featuring archaeological themes serve both as responses to and illustrations of the theories set out in two important paratexts relating to the discipline: the foreword of the *Légendes rustiques* and the preface to Laisnel de la Salle's *Croyances et légendes du cœur de la France*. The following is one of the most significant extracts from the former work:

> Cependant l'esprit gaulois a légué à toutes nos traditions rustiques de grands traits et une couleur qui se rencontrent dans toute la France, un mélange de terreur et d'ironie, une bizarrerie d'invention extraordinaire, jointe à un symbolisme naïf qui atteste le besoin du *vrai* moral au sein de la fantaisie délirante.
>
> Le Berry, couvert d'antiques débris des âges mystérieux, de tombelles, de dolmens, de menhirs et de *mardelles* (voyez pour ces mystérieux vestiges l'*Histoire du Berry*, par M. Raynal), semble avoir conservé, dans ses légendes, des souvenirs antérieurs au culte des druides: peut-être celui des dieux kabyres, que nos antiquaires placent avant l'apparition des Kymris sur notre sol. Les sacrifices de victimes humaines semblent planer, comme une horrible réminiscence, dans certaines visions. Les cadavres ambulants, les fantômes mutilés, les hommes sans tête, les bras et les jambes sans corps, peuplent nos landes et nos vieux chemins abandonnés.[25]

The presence of archaeological themes in George Sand's thought is therefore intimately linked to rustic legend and is intended to be put to the service of the "vrai moral", a notion to which she would return in the preface of *Jeanne*. Certain Berrichon legends could then be understood as the last remaining vestiges of Gallic culture, and the natives of Berry the direct inheritors of this ancient people.

Yet although Sand may have played the role of apprentice archaeologist herself, she did not see fit to introduce the figure of the archaeologist into any of her works. Instead it fell to her son, Maurice Sand, undoubtedly influenced by his mother's rich store of archaeological experiences, to write *Callirhoé*, a work of what may be termed archaeo-fiction, featuring an archaeologist in the role of hero. This novel, published by Michel Lévy in 1864, is set in a location not far from la Motte de Presle, at the heart of the Berry region. The nature and the sheer originality of this work is intimately linked to the young novelist's evocation of madness and of the fantastical, themes largely

[25] George Sand, *Légendes rustiques* (Paris: A. Morel et Cie, 1858), p. vi.

rooted in his mother's observations of the excavations of 1853.[26] By contrast, George Sand preferred to model characters who were carried away by their enthusiasm for archaeology. How then can we explain her creation of these dilettante archaeologists?

The key to this question is to be found in Sand's own writings, in *Le Péché de Monsieur Antoine*, in which she pours contempt upon a certain sub-section of archaeology's most enthusiastic practitioners:

> Qu'y a-t-il de plus ridiculement crédule et de plus facile à tromper qu'un pédant à idées préconçues? Je me souviens d'un antiquaire qui vint ici l'an passé: il voulait trouver des pierres druidiques, et il en voyait partout. Pour le satisfaire, je lui montrai une vieille pierre que des paysans avaient creusée pour y piler le froment dont ils font leur bouillie, et je lui persuadai que c'était l'urne où les sacrificateurs gaulois faisaient couler le sang humain. Il voulait absolument l'emporter pour la mettre dans le musée du département. Il prenait tous les abreuvoirs de granit qui servent aux bestiaux pour des sarcophages antiques. Voilà comment les plus ridicules erreurs se propagent. Il n'a tenu qu'à moi qu'une bâche ou un pilon passassent pour des monuments précieux. Et pourtant ce monsieur avait passé cinquante ans de sa vie à lire et à méditer.[27]

In short, then, the archaeologist is presented as a somewhat insipid character, lacking in literary depth. Nonetheless, in *La Famille de Germandre*, Sand gives us the character of Sylvain de Germandre, a gentleman whose passion for archaeology has led him back to the land. The forename "Sylvain" clearly signals his status as an "homme de campagne", an expression which is applied to him at numerous points in the text. This linkage of themes relating to archaeology and to rural life, undertaken with a view of making the quasi-mythical voice of the French peasantry heard, constitutes a topos that is characteristic of George Sand's work. The novel itself is a curious story revolving around the theme of inheritance. In order to become the heir of the Marquis de Germandre, the hero must open the casket guarded by the Sphinx. During a stroll with his cousin, the Chevalier de Germandre adopts the role of *cicerone*:

---

[26] Claire Le Guillou, "Du roman archéologique à l'archéologie du roman: *Callirhoé* de Maurice Sand", presented at the Conference "La plume et la pierre: L'écrivain et le modèle archéologique au XIXe siècle", Centre Universitaire de Formation et de Recherche de Nîmes, 3-6 July 2006. Conference proceedings in preparation.

[27] George Sand, *Le Péché de Monsieur Antoine* (Meylan: Editions de l'Aurore, 1982), pp. 281-82.

> Il se mit à parler archéologie pour changer de conversation. Il expliqua à madame de Sévigny une inscription latine dont un fragment à peine lisible apparaissait sur une pierre encastrée dans la muraille, et qui prouvait que la chapelle de Sainte-Denise avait été bâtie sur les ruines d'une chapelle dédiée à Dionys ou Dionysos, le dieu antique apporté dans les Gaules par la conquête romaine.
>
> Hortense s'étonna de lui voir déchiffrer cette inscription tracée en abrégé, et restituer non seulement les lettres retranchées, mais les mots entièrement détruits.
>
> - Où donc avez-vous appris tout ce que vous savez? lui dit-elle.
>
> - Oh! Ceci n'est rien, répliqua le chevalier en riant de lui-même avec bonhomie; j'en sais bien d'autres! Je suis une espèce de savant, moi, sans en avoir la mine! C'est un ridicule de plus que j'abandonne à la moquerie. Imaginez-vous que, vivant aux champs, où j'eusse dû, tout naturellement, m'occuper de botanique ou d'entomologie, d'une branche quelconque de l'histoire naturelle qui m'eût servi en agriculture, et dont les matériaux se trouvaient à foison sous ma main, je m'en suis allé donner tête baissée dans des études qui ne pouvaient profiter en aucune façon aux autres ni à moi-même dans la situation où je me trouvais.[28]

These archaeologically rich spaces serve as the backdrop to the protagonists' revelation of their love for one another. In this passage, which leads up to the revelation scene, Sand insists on the Chevalier's epigraphical talents, though these are not viewed as any real kind of distinction on the basis that archaeology is not a "useful" science. In any case, the Chevalier does not see himself as an archaeologist, but rather as a numismatist. Yet here, once again, the Chevalier shows great humility, particularly in his reflections on the conflict between his own intellectual ambitions and his material needs. "C'est amusant, mais ça prend bien du temps qui serait mieux employé, dit-on, à faire fortune. Que voulez-vous! Quand on ne sait pas s'enrichir! Vous voyez, c'est la science du passé, la science de la mort, que j'ai prise pour antithèse de la culture de mon coin de terre, où mieux vaudrait appliquer la science de la vie!" (pp. 194-95). His apparently pointless pastime will, however, prove to be the key to the novel, effectively the *deus ex machina* which will permit him to escape from poverty and find a better life. The Chevalier de Germandre successfully passes the trial of the Sphinx and, thanks to his skill as a "numismate exercé" (p. 290), he manages to open the famous casket. Archaeology is thus no longer seen as a "science de la mort", but rather as a science securely anchored in present-day life. Throughout

[28] George Sand, *La Famille de Germandre* (Paris: Michel Lévy, 1861), pp. 189-90. Subsequent references to this edition are given after quotations in the text.

her *œuvre*, Sand aimed to present archaeology and its associations as part of a living and idealistic process, free from the macabre connotations which had struck her so clearly in Rome.

If *La Famille de Germandre* has been largely ignored by Sand scholars, *Jeanne* and *Nanon* have, by contrast, aroused much more critical interest. In these novels, the presence of standing stones is not simply or exclusively an element of set-dressing, but serves rather to situate the characters in a quasi-mythical era, favourable to the establishment of the "vrai moral". In other respects, the novels seem almost to reflect one another across a distance of twenty-eight years of the author's life.[29] If the Celtic monuments, named "aire aux Fées" or "trou aux fades" may only be accessed with great difficulty, the former will ultimately serve as a kind of homeland for Jeanne, while Nanon will eventually be exiled to the latter. These two mystical spaces, constructed as havens of peace and revelation, lead one of the heroines to celibacy and death, and the other, by contrast, to marriage and fulfilment.

From the opening lines onward, it is clear that archaeology and the passion for all things Celtic so beloved of the Romantics will be dominant themes in the novel *Jeanne*. To the extent that archaeological relics remain hidden and the Jomâtre Stones are initially almost invisible in the text, the author establishes a need for them to be in some way rediscovered. The narrator makes them visible in drawing them to the reader's attention, even taking the trouble to point out the road which leads to them.[30] These initially invisible locations thus permit Sand to retrace a story forgotten by official history and its practitioners long ago, in recounting the tale of a disinherited people and their homeland, of which the geographical and political outlines are so ill-defined.

In the prologue, Sand compares Jeanne to a "druidesse endormie" and more specifically to Velléda (pp. 37-38),[31] thus setting

[29] Brigitte Lane, for her part, makes the following remark: "*Nanon* marque non seulement un tournant majeur dans l'œuvre de George Sand, mais l'aboutissement de la 'quête' idéologique et esthétique à laquelle l'écrivaine s'est livrée pendant près de trente ans, depuis *Jeanne*." See "George Sand, 'ethnographe' et utopiste", *Revue des Sciences Humaines*, avril-juin 1992, n°226, p. 156.

[30] George Sand, *Jeanne* (Meylan: Editions de l'Aurore, 1986), pp. 33-34. Subsequent references to this edition are given after quotations in the text.

[31] See Ione Crummy, "Les réincarnations de la druidesse Velléda", *George Sand et l'écriture* (Montréal: Université de Montréal, 1996), pp. 405-14.

out the tragic destiny of her young heroine and insisting on the fact that "nous croirions plutôt que le pur sang de la race gauloise primitive s'est conservé jusqu'à nos jours sans mélange, dans quelques tribus rustiques de nos provinces centrales" (p. 38). Jeanne's Gallic lineage is all the more clearly indicated by the fact that she is the daughter of Tula, the feminized form of Tullum, which was the Latin name of the town of Toulx. Sand thus lends a mythical dimension to her peasant figures. In order to consolidate this theme, Sand places great emphasis on the archaeological backdrops. Thus, in the first chapter, entitled "La ville gauloise", the author invites us to continue our archaeological visit around Toulx-Sainte-Croix in the company of Guillaume de Boussac, who is visibly enthused by the discipline. Sand then describes the town, showing off both her own knowledge of history and archaeology, and that of the town-dwellers (pp. 43-45, 47). Then, in Chapter V, it is the turn of the village priest to reveal his archaeological erudition and to explain the region's Gallic history, in comparing the site of the Jomâtre stones, once consecrated to the bloodthirsty cults, and the stones at Ep-Nelle, a safe and beneficent place. The stong insistence on the Gallic origins of this town and the simultaneous refusal of any Roman contribution to its founding serves as a means of contrasting it with Boussac, a modern town which will become a prison for Jeanne in which she will finally end her life.

In *Nanon*, the Celtic monuments of Crevant are also described as a safe place, in direct contradiction of their modern role as a site of barbarism associated with the Revolutionary Terror. *Nanon* is a revolutionary novel, written in 1871, in the midst of another turbulent period of French history. The novel recounts the story of a peasant-girl and a young nobleman caught up in the torments of the Revolution and the Terror. Destined, despite his republican sentiments, to be arrested and condemned, Emilien de Franqueville is assisted in his attempts to escape and find a safe hiding-place by the young Nanon. It is at a crucial point in the novel, taking place in his refuge in the village of Crevant, that Sand chooses to bring archaeological themes into play. In the midst of this period of torment, it will be archaeology that permits the two beleagured young people to find their way to safety:

Nous savions qu'il y avait une ancienne voie romaine qui allait dans la direction du sud-est et nous n'avions pas d'étoile pour nous guider. Enfin, le ciel s'éclaircit et nous vîmes au-dessus des arbres la Ceinture-d'Orion, que les paysans appellent les

> Trois-Rois. Dès lors, nous trouvâmes la voie sans peine. Elle était bien reconnaissable à ses grosses rainures de pierres sur champ.[32]

These first traces and their knowledge of archaeology offer them salvation, with the Roman road leading them to the Gallic ruins at Crevant, which are not initially visible. Sand thus leads us backward through time, in order to take us to the quasi-mythical origins of Berry. They arrive safe and sound in an "oasis de granit et de verdure, un labyrinthe où tout était refuge et mystère". Simone Bernard-Griffiths talks of a "lieu sacré", and of the "insularité temporelle paradoxale en pleine Terreur".[33]

They then set up home in a quarrier's cottage, which has in fact been constructed around an ancient dolmen. Sand thus institutes temporal continuity between the Celtic era and the late eighteenth century in which the novel takes place, by introducing the theme of recycling and the reformulation of language, for the site is thereafter renamed "l'aire aux fées". She reinforces the line of temporal continuity by presenting the Berrichon peasants once again as the last vestiges of the Celtic people (p. 164), emphasizing the fact that the "aire aux fées", having been occupied by "les *femmes sauvages* (les druidesses), avait servi d'ermitage à des saints et à des saintes" (p. 150). Sand also makes nature participate in this temporal continuity, conferring the status of archaeological relic upon the trees, "d'un âge incalculable", "[qui] peuvent présenter un spécimen de la Gaule primitive dans son intégralité" (p. 165). In this place which is so totally steeped in archaeology, Sand describes the contentment of the peasant-girl Nanon and the young aristocrat Emilien de Franqueville, thus delimiting a space which is more than a mere utopia, but which may rather be understood as an *uchronie*, taking place outside of history as we know it.

However, as in the novel *Jeanne*, Sand makes a clear distinction between two types of Celtic monument. The "parelle" is a "mauvaise pierre", "signe d'une frontière possible entre naturel et surnaturel", and the Druiderin a place of sentimental revelations. It is at the Druiderin, a site which is barely visible and which is also much

---

[32] George Sand, *Nanon* (Meylan: Editions de l'Aurore, 1987), p. 140. Subsequent references to this edition are given after quotations in the text.

[33] Simone Bernard-Griffths, "L'Espace dans *Nanon*: de la géographie à la mythologie", in *George Sand et l'écriture* (Montréal: Université de Montréal, Bibliothèque nationale du Québec, 1996), pp. 339-53.

favoured by Emilien, that old Dumont will tell Nanon that Emilien is in love with her (p. 175). By such means, George Sand contrives to enrich the places she describes. For, in her opinion, archaeology serves to reveal that the relics taken from the earth are neither dead nor unmodifiable. They are rather an element of social reconciliation, as when, for example, peasantry and aristocracy end up by being united, as in *Nanon* and in *La Famille de Germandre*. They are, in short, a portal leading the happy few towards fortune and happiness, and as such, they constitute a return to a kind of golden age. In spite of these positive associations, however, Sand on numerous occasions evokes the partial, or even total, destruction of archaeological relics that she so admires:

> Le grand *Durderin* (corruption de *Druiderin*) est encore debout et l'ensemble de l'île aux Fades n'a pas trop changé; mais elle a perdu son nom, les fées se sont envolées et le voyageur qui chercherait leur ancien séjour serait forcé de demander à la ferme voisine du Petit-Pommier, le chemin des Grosses-Pierres. Moins de poésie à présent, mais plus de travail et moins de superstition. (p. 165)

Yet although Sand repeatedly emphasizes in *Nanon* the extent to which the modern world has been rendered prosaic, the ultimate corruption of the site in fact gives rise to few real regrets. This becomes all the more evident in "Le Marteau rouge", in which Sand, who shows once again her tremendous wealth of erudition, recounts the history of an archaeological relic, from its creation to its final destruction. In the final analysis, then, the author's use of archaeology facilitates the presentation of the steady evolution of the world, which is depicted as an entity that is constantly in a state of becoming, and which is subjected irrevocably not only to the whims of mankind, but also and most especially, to the power of Nature.

The closing lines of this story, which represent the last trace of interest in archaeology to be found in Sand's corpus of writings, may therefore be seen to offer the perfect conclusion to our present deliberations:

> Tel est le sort des choses. Elles n'existent que par le prix que nous y attachons, elles n'ont point d'âme qui les fasse renaître, elles deviennent poussière; mais, sous cette forme, tout ce qui possède la vie les utilise encore. La vie se sert de tout, et ce que le temps et l'homme détruisent renaît sous des formes nouvelles, grâce à cette fée qui

ne laisse rien perdre, qui répare tout et qui recommence tout ce qui est défait. Cette reine des fées, vous la connaissez fort bien: c'est la nature.[34]

Translated by Louise Lyle

[34] George Sand, "Le Marteau rouge", *Contes d'une grand-mère*, Deuxième série, (Meylan: Éditions de l'Aurore, 1983), p. 150.

SECTION 3

LATE NINETEENTH CENTURY

Jules Verne and the Discovery of the Natural World

Tim Unwin

Abstract: *This chapter explores the elemental forces at work in Verne's vast series of* Voyages extraordinaires. *It reveals the extent to which the author's representation of nature's extreme environments reflected a contemporary literary preoccupation with the pedagogical exposition of new geographical, anthropological, geological, zoological, botanical and technological knowledge, and the stylistic processes of linguistic accumulation, appropriation and assimilation used to achieve this.*

Few fictional undertakings are as ambitious or as wide-ranging as Jules Verne's *Voyages extraordinaires*. Few are so quintessentially of their time. In mapping out the nineteenth century's new understanding of the globe and its relationship with it, Verne gives a uniquely valuable insight not only into the state of science and technology, but also of the discovery of the natural world that modern progress has enabled. Throughout his colossal enterprise Verne aims to provide systematic and comprehensive coverage of the globe, while also putting together a compendium of current knowledge about it through the texts and documents he so conspicuously uses in the making of his stories. Perhaps inevitably, the aim of totality itself turns out to be a vast fiction. However Verne, in the encyclopaedic spirit of his time, manages to offer an astonishingly wide-ranging view of the century's discoveries and developments: geographical, anthropological, geological, zoological, botanical, technological and so on. In terms of geographical coverage, he maps out the continents, the seas and the polar regions, returning obsessively to key points of the globe in several novels. The four elements – earth, air, fire and water (or its solid equivalent, ice) – associated with different aspects of the globe, mark out the parameters of the *Voyages extraordinaires*, and give them a mythical scope that has produced a rich and continuous strand of scholarship over recent decades.[1]

In the following pages, my focus will be not only on Verne's discourses about nature, but also – and crucially – on the nature of his discourses. How is nature understood and mediated in the *Voyages extraordinaires*? What language and idioms does the writer have at his disposal? How do the discoveries of recent explorers affect the writing of novels? And how might Verne's writing of nature be said to differ from that of some of his contemporaries or near-contemporaries? While I shall duly stress Verne's expansive and cumulative manner of writing, I should also insist at the outset on the pedagogical remit of his fiction, which sets out specifically to educate the young not only by initiating them into the scientific and technological developments of the time, but also by giving them an overview of the nineteenth-century global village through long encyclopaedic interpolations. However, my central point is that, for all the knowledge that his stories relay and exploit, Verne is essentially a self-conscious writer who uses the century's new-found understanding of the natural world as a means to experiment with the language of fiction itself. Scientific and other discourses collide and collude throughout his work, in a manner that specifically draws attention to their textual status. And journeys, with their apparently linear progression interrupted by digressions or obstacles, provide not only the structure and metaphor, but also the very text of his narratives since they are so clearly negotiated in and through documents of all kinds – diaries, logbooks, guidebooks, manuals, dictionaries, encyclopaedias, newspapers and so on. In this sense, Verne emerges not just as a great scientific popularizer and a pedagogue, but as an experimental writer "pushing back the frontiers"

[1] This strand of scholarship was apparent as early as 1949 in a groundbreaking article by Michel Butor, "Le Point suprême et l'âge d'or à travers quelques œuvres de Jules Verne", in *Répertoire I* (Paris: Minuit, 1960 [1949]), pp. 130-62. More recently it has been exploited most notably by Simone Vierne in *Jules Verne. Mythe et modernité* (Paris: Presses Universitaires de France, 1989). A productive application of this approach to a single text, *Les Indes noires*, can be seen in David Meakin, "Future Past: Myth, Inversion and Regression in Verne's Underground Utopia", in E. J. Smyth (ed.), *Jules Verne: Narratives of Modernity* (Liverpool: Liverpool University Press, 2000), pp. 94-108.

of fiction through his extension of it into new and ostensibly unfictional discourses. In so doing, he provides us not only with a broad series of "histoires de la terre", but also gives us "la terre en histoires", literally turning the globe into text and fiction.

Before homing in on the *Voyages extraordinaires*, though, and by way of contextualizing Verne's approach, I want to refer to two other, slightly different examples of nineteenth-century novelists' uses of the "discovery" of the natural world. The first, Balzac's eulogy of Cuvier from the opening stages of *La Peau de chagrin*, represents an early incursion of palaeontology into fiction; the second is a pantheistic description of nature in Flaubert's 1842 novel *Novembre* (one of several such passages in his early writings). As I shall argue, both writers use nature, in their very different ways, specifically to position themselves as artists and novelists.

This is the passage from Balzac:

> Vous êtes-vous jamais lancé dans l'immensité de l'espace et du temps, en lisant les œuvres géologiques de Cuvier? Emporté par son génie, avez-vous plané sur l'abîme sans bornes du passé, comme soutenu par la main d'un enchanteur? En découvrant de tranche en tranche, de couche en couche, sous les carrières de Montmartre ou dans les schistes de l'Oural, ces animaux dont les dépouilles fossilisées appartiennent à des civilisations antédiluviennes, l'âme est effrayée d'entrevoir des milliards d'années, des millions de peuples que la faible mémoire humaine, que l'indestructible tradition divine ont oubliés et dont la cendre, entassée à la surface de notre globe, y forme les deux pieds de terre qui nous donnent du pain et des fleurs. Cuvier n'est-il pas le plus grand poète de notre siècle? [...] Il réveille le néant sans prononcer des paroles artificiellement magiques, il fouille une parcelle de gypse, y aperçoit une empreinte, et vous crie: "Voyez!" Soudain les marbres s'animalisent, la mort se vivifie, le monde se déroule![2]

And this is the passage from Flaubert:

> L'esprit de Dieu me remplissait, je me sentais le cœur grand, j'adorais quelque chose d'un étrange mouvement, j'aurais voulu m'absorber dans la lumière du soleil et me perdre dans cette immensité d'azur, avec l'odeur qui s'élevait de la surface des flots; et je fus pris alors d'une joie insensée, et je me mis à marcher comme si tout le bonheur des cieux m'était entré dans l'âme. [...] Et je compris alors tout le bonheur de la création et toute la joie que Dieu y a placée pour l'homme; la nature m'apparut belle comme une harmonie complète, que l'extase seule doit entendre; quelque chose de tendre comme un amour et de pur comme la prière s'éleva pour moi du

---

2 Honoré de Balzac, *La Peau de chagrin*, in *La Comédie humaine*, 11 vols (Paris: Gallimard, 1979), X, p. 74-75.

> fond de l'horizon, s'abattit de la cime des rocs déchirés, du haut des cieux; il se forma, du bruit de l'Océan, de la lumière du jour, quelque chose d'exquis que je m'appropriai comme d'un domaine céleste, je m'y sentis vivre heureux et grand, comme l'aigle qui regarde le soleil et monte dans ses rayons.[3]

Now it is not hard to find ways in which these passages stand in contrast to one another. While Balzac talks dithyrambically of prehistory, Flaubert is focussed mystically on something much nearer the present. Balzac is (ostensibly, at least) speaking of one of the nineteenth century's first great scientific icons, while Flaubert appears to be offering an unmediated, and quite unscientific description of a personal experience. Balzac conveys his vision through an explicit reference to scientific discourse, while Flaubert's is intensely lyrical. And if Flaubert's framework is one of pastoral harmony, far removed from the city and signs of civilization, Balzac evokes, alongside the natural world, the spectre of past and even present civilizations that, like nature itself, come and go. Most significantly, while Balzac's vision is of a creative act that restores plenitude and meaning almost *ex nihilo*, Flaubert locates and verbalizes an already existing abundance of meaning in the natural scene around him. Balzac, it should be added, writes at a time when the story of dinosaurs is just beginning to emerge fully – heralding an era of fieldwork and discovery in the 1840s that has been comically described as "la ruée vers l'os" – whereas Flaubert evokes a natural world that, for him, is already complete, and needs only to be observed and transcribed in its dramatic presence and fullness.

For both, though – and this is the crucial point – the discovery of nature is absolutely synonymous with that of the power of language. Despite many differences between them, each articulates through the perception of nature a vision and a poetics that will be central to their approach as novelists. Faced with the spectacle of nature at first or at second hand, ancient or modern, scientifically mediated or mystically intuited, the writer finds in it a source of rich significance and of analogies with his own creative processes. Balzac explicitly builds his passage on the notion that observed reality would be nothing, literally remaining in almost total obscurity, without the

[3] Gustave Flaubert, *Novembre*, in *Œuvres de jeunesse* (Paris: Gallimard, 2001), p. 781.

poetic intervention of creative geniuses such as Cuvier – and thus stakes out his own position as an artist whose task is akin to that of the palaeontologist, reconstructing the spectacle of totality from the discovery of fragments. In Cuvier's creative interrogation of distant geological eras, Balzac finds his cue as the secretary of his own contemporary society, and he articulates an aesthetic of accumulation and totalization. And here is the link with Flaubert, since nature is for Flaubert too the intuition of totality. Nature, in Flaubert's experience, may need no reconstruction as it does for Balzac or, beyond him, Cuvier, but it is precisely in this wholeness and abundance that the writer finds his own way forward. If Flaubert does not explicitly outline an aesthetic approach in the passage quoted, the discovery of the natural world is clearly enhanced by the process of writing itself. Flaubert turns a raw mystical experience (if indeed such a thing ever took place) into a quintessentially written one. And, as in the case of Balzac, but for very different reasons, the writing of that vision of nature progressively gathers momentum and speed. Words multiply and acquire an intensity that puts the primary emphasis on their rhythmic and phonetic rather than their referential status, until it seems that language itself becomes the real locus, and perhaps the real focus, of the experience referred to. Nature, we might conclude, must be viewed and enhanced through language in order to be fully seen, appreciated and understood. Much more than the point at which a mystical experience is retrospectively re-enacted, writing is the basis of such an experience.

So, in these works that mark an early stage of each of the two novelists' development, an aesthetic credo is at some level being set out. The discovery of the natural world, or the representation of that discovery, implies a view of writing as an act of creation, completion and totalization. The puzzling fragmentedness or the abundant diffusion of nature is transformed through the act of writing into a vision of coherence, wholeness and, most strongly in Flaubert's case, harmony. Now, much has of course been said and written about disruptions and discontinuities of meaning in both Balzac and Flaubert – notwithstanding the fact that each of them appears to predicate his approach on notions of unity and order – and it is not my intention here to go back to such discussions. If I have chosen, however, to draw attention to what I shall call their "poetic recuperation" of nature, it is because at the outset it offers us a major point of contrast with Jules Verne (there will also ultimately be significant points of similarity, to which I shall return). Verne's writing is not underpinned by anything like the same systematic discourse of aesthetic unity or harmony as Balzac's or Flaubert's. Writing is not presented as an undertaking which aims to create formal structural coherence or indeed patterning out of chaos. The writing of nature is, in the first instance, a process of extensive and detailed transcription. There is, indeed, rarely any explicit or formal aesthetic discussion in Verne, who does not align himself with the poet or the mystic, and who bypasses those philosophical discourses that are so evident in writers like Balzac and Flaubert. And while there is very clearly a process of mesmerizing accumulation in the *Voyages extraordinaires*, just as there seems to be in those passages from Balzac and Flaubert, this is most frequently geared towards fragmentation, diffusion, dispersion and multiplication, rather than towards unification or harmonization. For Verne, there are apparently no reductive procedures that can account for and structure the infinite variety and complexity of the world. Rather than synthesis, his fundamental impulse is towards mathesis:[4] Verne always prefers the inventory conducted *in extenso* – hence the massive lists of flora and fauna in works such as *Vingt mille lieues sous les mers*. These literally give the impression that the natural world is infinite, and – similarly – that the language used to describe it is unbounded and ever equal to the task, since there is always another term available to the writer to detail another phenomenon. The dictionary matches nature with its endless variety and largesse, yet this near-perfect fit between the natural world and the textual one, far from producing a sense of wholeness or plenitude, leads on many occasions to an effect of arbitrariness. The unknown world is turned into words that are, equally, unknown to all but the absolute specialist, and since so many different specialisms come into play in Verne's work, the ideal reader would have to be a polymath to make full sense of the text. But making sense is not really the point. Often, meaning beyond the sounds and the strangeness of

[4] "The *Voyages*, anticipating the closure of the circle of knowledge, the attainment of omniscience, aim to supplant mimesis by mathesis, deploying science to abolish fiction. But the Vernian savant is always threatened, in the course of his journeys of intellectual discovery, by the catastrophic reduction of science to nescience" (Andrew Martin, *The Knowledge of Ignorance: From Genesis to Jules Verne*, (Cambridge: CUP, 1985), pp. 6-7).

the words themselves seems to disappear, as in this list, one among many other possible examples of this distinctively Vernian process:

> Pendant deux heures toute une armée aquatique fit escorte au *Nautilus*. Au milieu de leurs jeux, de leurs bonds, tandis qu'ils rivalisaient de beauté, d'éclat et de vitesse, je distinguai le labre vert, le mulle barberin, marqué d'une double raie noire, le gobie éléotre, à caudale arrondie, blanc de couleur et tacheté de violet sur le dos, le scombre japonais, admirable maquereau de ces mers, au corps bleu et à la tête argentée, de brillants azurors dont le nom seul emporte toute description, des spares rayés, aux nageoires variées de bleu et de jaune, des spares fascés, relevés d'une bande noire sur leur caudale, des spares zonéphores élégamment corsetés dans leurs six ceintures, des aulostones, véritables bouches en flûte ou bécasses de mer, dont quelques échantillons atteignaient une longueur d'un mètre, des salamandres du Japon, des murènes échidnées, longs serpents de six pieds, aux yeux vifs et petits, et à la vaste bouche hérissée de dents, etc.[5]

The conclusion of the list with "etc.", a frequent device in the *Voyages extraordinaires*, reinforces the notion that, in an unbounded linguistic and/or natural world, the writer must arbitrarily halt his descriptions. There is no necessary stopping point, for there *is* nothing that divides, segments or arrests the continuum. It is easy to see, from such lists, why translators of Verne have such immense difficulty: it is like translating the encyclopaedia or the dictionary, and often the work involves retracing Verne's own steps back into the reference works, only to find that the terminology is still missing or indeed that Verne himself may have invented a likely name.[6] In such passages, Verne celebrates the nineteenth century's invention of new languages and terminologies, but he often seems to do so in a conspicuous vacuum of significance. And while Verne's lists have sometimes been judged to have a poetic, incantatory force – a point I shall be returning to – their only obvious function as discourse is the endeavour to reference every variety of every species. This mirrors the functioning of his writing more generally, in its attempt to map out the totality of the globe itself – its seas, mountains, jungles, rivers, ice caps – in words. It seems that the writer must simply get out there and attempt to cover everything in text, literally throwing words at everything, following and inventory-

---

[5] *Vingt mille lieues sous les mers*, p. 153. References to Jules Verne's novels are to the fifty-volume *Les Œuvres de Jules Verne* (Lausanne: Rencontre, 1966-71).

[6] See for example the remarks by William Butcher, "Note on the Text and Translation", in Jules Verne, *Twenty Thousand Leagues Under the Seas* (Oxford: Oxford University Press, "World's Classics", 1998), pp. xxxvii-xxxviii.

ing every contour and every particularity of nature. Thus, as has sometimes been argued, writing is also for Verne a profoundly imperialistic pursuit, in its attempt at appropriation, coverage and possession through language.[7] In *Le Tour du monde en quatre-vingts jours*, we are told that Her Majesty's dominion extends to the remotest corners of the Indian sub-continent. So too, more generally, the Vernian text tracks the meanders of every journey and every exploration, spreading its imperialistic lexis, and checking off every location-turned-text.

So style, in the early Balzacian or Flaubertian sense of a unifying or harmonizing poetics that is continuous with, and can be inferred from, the processes of writing, is apparently not part of Verne's vision of and for the novel. Not that style, in the sense of pure verbal elegance and linguistic propriety, is unimportant to him. On the contrary, he attaches explicit value to it throughout his career as a novelist, and there are many finely written passages in the *Voyages extraordinaires*.[8] However, there is no discourse in Verne that suggests he might be making broader claims for his contribution to the genre, and he does not appear to build into his work any implicit or explicit philosophical or aesthetic propositions about it. The frequently held view that Verne was a literary outsider seems also to have much to do with the way he positions himself in relation to his reading public, or rather by the way he is positioned by his editor Hetzel. From the outset in 1863, when Hetzel publishes Verne's first novel, *Cinq semaines en ballon*, the remit he hands down to his protégé is that of the educator, targeting the younger reader with a mixture of science, geography and adventure. It is the classic mix of "plaire et instruire", though the level of instruction is predominantly

---

[7] See for example Andrew Martin, *The Mask of the Prophet. The Extraordinary Fictions of Jules Verne* (Oxford: Clarendon, 1990), pp. 19-25.

[8] When writing *Voyages et aventures du capitaine Hatteras* in 1864, Verne engages in some discussion with Hetzel about how to present certain aspects of his story, and adds a remark about his own literary ambitions: "Tout ceci, c'est pour vous dire combien je cherche à devenir un *styliste*, mais sérieux; c'est l'idée de toute ma vie" (Letter to Hetzel of 25 April 1864, in, *Correspondance inédite de Jules Verne et de Pierre-Jules Hetzel (1863-1886)*, ed. by Dumas, Olivier, Piero Gondolo della Riva et Volker Dehs, 2 vols (Geneva: Slatkine, 1999-2001), I, p. 28). Verne himself was later to regret that he had not been taken seriously as a literary stylist: "Je ne compte pas dans la littérature française", he told an interviewer. See R.H. Sherard, "Jules Verne at Home: His Own Account of his Life and Work", *McClure's Magazine* (January 1894), online at: http://jv.gilead.org.il/sherard.html.

factual and concrete rather than, say, moral, religious or philosophical. Arthur Evans argues that the Loi Falloux of 1850, giving a large hand to Catholicism in education, had halted the progress of the curriculum towards the inclusion of science.[9] With the 1866 launch of his "Bibliothèque d'Education et de Récréation", Hetzel targets a gap that will later be filled by secular scientific education, and Jules Verne is naturally his flagship author. In his 1866 preface to *Voyages et aventures du capitaine Hatteras*, Hetzel announces that the forthcoming series of novels by Jules Verne will provide a compendium of all current and available knowledge. Thus he is drawn to formulate that totalizing ambition which will remain at the heart of the *Voyages extraordinaires*: "Son but est, en effet, de résumer toutes les connaissances *géographiques, géologiques, physiques, astronomiques*, amassées par la science moderne, et de refaire, sous la forme attrayante et pittoresque qui lui est propre, l'histoire de l'univers".[10] Totalization, it should be added, is here to be understood as the open-ended accumulation of knowledge in breadth. Verne's fictional enterprise offers an overview of what Claude Bernard called "l'état actuel de nos connaissances",[11] but this is necessarily provisional, and by definition it cannot capture that underlying, synthesizing or harmonizing sense of totality that authors like Balzac and Flaubert so clearly envision.

Hetzel's early statement thus sets the agenda, defining the scope and the ambition of Verne's project and indeed the very nature of his literary ambition. Total coverage of the globe and of our knowledge about it will remain as one of the defining characteristics of the *Voyages extraordinaires*, the guiding impulse that shapes the author's vision and our reading of his work. As Michel Serres points out, Verne's work thus

---

[9] See Arthur B. Evans, *Jules Verne Rediscovered. Didacticism and the Scientific Novel* (Westport, Connecticut: Greenwood Press, 1988), pp. 9-15.

[10] P.-J. Hetzel, preface to *Voyages et aventures du capitaine Hatteras* (Paris: Hetzel, 1866), p. 2.

[11] "Quand nous faisons une théorie, générale dans nos sciences, la seule chose dont nous soyons certains, c'est que toutes ces théories sont fausses absolument parlant. Elles ne sont que des vérités partielles et provisoires qui nous sont nécessaires, comme des degrés sur lesquels nous nous reposons, pour avancer dans l'investigation; elles ne représentent que l'état actuel de nos connaissances, et, par conséquent, elles devront se modifier avec l'accroissement de la science, et d'autant plus souvent que les sciences sont moins avancées dans leur évolution. See Claude Bernard, *Introduction à l'étude de la médecine expérimentale* (1865), UQAC online edition consulted September 2007, p. 41: http://classiques.uqac.ca/classiques.

represents a cartography of knowledge itself, a modern epic of Homeric proportions.[12] Julien Gracq notes the same essential impulse, while focussing on Verne's geographical coverage. At the heart of the corpus, he says, are the "grands romans cosmiques" based on one of the fundamental elements – the sea (*Vingt mille lieues sous les mers*), the air (*Cinq semaines en ballon*), the earth (*Voyage au centre de la terre*) – or on one of the key ice-bound points of the globe such as the North Pole (*Voyages et aventures du capitaine Hatteras*), or the Antarctic (*Le Sphinx des glaces*). Then there is a whole series of novels that are set in key points of the various continents. And finally, says Gracq, there are the novels that link different regions and continents, binding the Vernian corpus into a coherent entity, such as *Le Tour du monde en quatre-vingts jours*, or *Les Enfants du capitaine Grant*.[13]

Serres and Gracq rightly emphasize the very self-conscious attempt at totality in Verne, both of knowledge itself and of the natural world that is the outward focus of such knowledge. One could additionally stress that the discipline of geography plays the pre-eminent role in that process. Geography is, first and foremost – and self-evidently – the discipline most closely aligned to travel, involving knowledge about conditions and features of different parts of the natural world. However, geography is also the key that unlocks a multitude of other knowledge areas in the *Voyages extraordinaires*. First among them are oceanography, geology, volcanology, cartography, botany, zoology and anthropology. Following on from these (when Verne's travellers enter into the detailed calculation of distances and global positions) we find astronomy, geometry and trigonometry, or (when certain modern forms of travel are involved) engineering, thermodynamics and aerodynamics. Conversely, when a text like *Voyage au centre de la terre* leads travellers towards the discovery of prehistory, we encounter archaeology and palaeontology. Last but by no means least, geography (as the study of places and peoples) is almost unfailingly coupled with history in Verne's novels, forming a continuum that remains at the heart of the French school curriculum to this day. Geography and the various areas of knowledge it leads out into thus facilitates Verne's encyclopaedic and "interdisciplinary" approach in the novel. Moreover, since these related fields are conspicu-

[12] Michel Serres, *Jouvences sur Jules Verne* (Paris: Editions de Minuit, 1974), pp. 11-17.
[13] "Entretien inédit: Julien Gracq", *Revue Jules Verne*, 10 (2000), 60-61.

ously referred to through their texts, often cited directly and referred to by their author, one of the most striking consequences of Jules Verne's work is the sheer quantity of text that has been recycled from other sources. As his travellers enter into new and different regions of the globe, Verne openly interpolates the accounts of historians, explorers, scientists and others, thus making of his fiction an amalgam and a hybrid of other people's accounts, and indeed offering a kind of "manual of manuals" within the fictionalized framework. The *Voyages extraordinaires* give us the whole world, but they also appear to give us the whole of knowledge about the world. Verne's travellers are often seen to venture into new terrains with book or manual in hand, or alternatively with remembered texts in their mind (for many of them have prodigious powers of recall). Authors and explorers are cited in abundance and often at random, nowhere more typically than in Verne's first major bestseller, *Cinq semaines en ballon*, where the hero, Dr Samuel Fergusson, patches together long tracts of textbooks and manuals as he speaks about the tribes, customs and history of the places that the balloonists visit. Crucially, there is no attempt to conceal these textual sources or to integrate them seamlessly into the new text. Even where authors are not named, the changes of style, tone and vocabulary provide sometimes as bumpy a ride for the reader as that journey in the balloon for Verne's travellers.

Given that part of Verne's approach involves the creation of a textual patchwork, it is not surprising that accusations of plagiarism surfaced regularly during his lifetime, and many such cases have been documented by his biographers. And his legacy too has often been overshadowed by references to the derivative nature of his writing. Sartre, in an amusing passage from *Les Mots* in which he comments on his own early uses of the Vernian principle of digression, tells how at key moments in the narrative he learned to incorporate encyclopaedia entries in order to pad out his stories.[14] That this derivative approach might be viewed in a positive light – as a daringly creative remapping not only of the world and of current knowledge but also of the novel and its discourses – is an argument that I should like to pursue more fully in the rest of this chapter.

Hetzel, we noted, was keen to promote the knowledge agenda in his early preface to *Hatteras*, insisting that Verne's novels would

[14] Jean-Paul Sartre, *Les Mots* (Paris: Gallimard, 1964), pp. 118-19.

provide good, wholesome instruction to the young at a time when science had become fashionable. What is less often noticed about this famous preface, though, is that Hetzel explicitly claims the stylistic high ground for his protégé. He points out that Verne has founded a new form of novel in which science, knowledge and discovery have fully entered into the purview of literature: "L'art pour l'art ne suffit plus à notre époque," he writes, "et [...] l'heure est venue où la science a sa place faite dans le domaine de la littérature". Crucially, he insists that this combination of science and of fiction has opened up an entirely new territory, which Jules Verne is able to exploit with true originality:

> Le mérite de M. Jules Verne, c'est d'avoir le premier et en maître, mis le pied sur cette terre nouvelle, c'est d'avoir mérité qu'un illustre savant, parlant des livres que nous publions, en ait pu dire sans flatterie: "Ces romans qui vous amuseront comme les meilleurs d'Alexandre Dumas, vous instruiront comme les livres de François Arago."

Now Hetzel's promotional brief here is obvious, and it might be unwise to read it entirely without scepticism. However, what his remarks suggest is that Jules Verne's approach creates a new hybrid genre, bringing together discourses that until then might have been considered quite separate. In this sense, Hetzel's preface can also be considered a key transitional moment in the nineteenth century, the moment at which the novel as a genre is seen and recognized to be breaking free of its attachment to a certain traditional notion of style. Hybridity – with its refusal to submit to notions of order, regularity, balance or harmony that are the classical tradition of rhetoric – is an aspect of Verne's work that has been most fully and most eloquently explored by Daniel Compère, who in *Jules Verne écrivain* points to the polyphonic nature of the Vernian text, constructed out of multiple voices and discourses which fuse together in a rich, challenging but intensely original mixture. Compère argues strongly that, for all their patchwork quality, and indeed largely because of it, Verne's texts nonetheless retain the mark of an individual voice.[15] So while, as I suggested earlier, Verne may not build into his texts an explicit reflection on the fabrication of text or an overt claim to be pushing the novel as a genre in new

[15] "Là est sans doute le paradoxe de ce texte fortement intertextuel: la présence des propos des autres donne au texte vernien une richesse, une puissance bien plus grande qu'à celui qui n'est composé que d'une voix isolée" (Daniel Compère, *Jules Verne écrivain* (Geneva: Droz, 1991), p. 87).

directions, there is in his approach an astonishing daring that in his own time is matched only by the later Flaubert with *Bouvard et Pécuchet*. That text, like Verne's *Voyages extraordinaires*, builds its own voices specifically through the uses and abuses of other texts that it displays, distorts or appropriates, and it too experiments with plurality rather than unity. It is premised on a fracturing principle that is radically different from that of early Flaubertian texts like *Novembre*. Running the constant risk of disappearing under the burden of other texts – that is part of its ironic and potentially self-destructive point – *Bouvard et Pécuchet* navigates a complex and treacherous path between them, extending novelistic discourses into every domain of human knowledge, and conversely allowing every domain of human knowledge to invade and perhaps to undermine the traditional terrains of novelistic rhetoric. Here, I would suggest, we find a striking similarity with Jules Verne who, for all the obvious popularity of his approach as an educational novelist targeting a mainly youthful readership, puts new languages and new terminologies on display in a fictional context, thus questioning and problematizing the very form of the novel. Add to this the highly self-conscious manipulations throughout the *Voyages extraordinaires* of literary devices and conventions (in particular, Verne's frequent experiments with the *robinsonnade*), the theatricality, the word-play, the narrativization of reading and writing themselves within Verne's plots, and a very different picture of his approach and contribution to the novel begins to build up.

George Orwell, echoing a long and dismissive tradition, once regretted that Verne should have been "so unliterary a writer", though he noted with some incomprehension that Verne appeared nonetheless to have the standard credentials and background of a nineteenth-century novelist.[16] The point we might make in response, with only minimal risk of paradox, is that Verne is literary precisely *because* he is so apparently unliterary, *because* he pushes the novel outwards towards those many other discourses that his century's scientists and explorers had produced. Yet it is also important to stress that this is not done at the cost or to the exclusion of those discourses which are more traditionally associated with the novel. There are from time to

[16] George Orwell, "Two Glimpses of the Moon", *The New Statesman* (18 January 1941), reprinted in *The Jules Verne Companion* (London: Souvenir Press, 1978), pp. 17-19.

time moments of intense poetic effusion in the *Voyages extraordinaires*, comparable with the passage from Flaubert's *Novembre* quoted earlier, and in which the writer, either through the discourses of science, or through the direct evocation of nature, has a predominantly aesthetic response to the phenomena he reports. As Michel Butor argues, his style is at times imbued with a poetic power where the richness and strangeness of words themselves becomes a source of fascination.[17] So there are also in Verne's work some astonishing coincidences of style and preoccupation with Balzac and Flaubert, despite the differences to which I drew attention earlier. Verne is perhaps closest to Balzac in certain pages of *Voyage au centre de la terre*, in his vivid and enthusiastic descriptions of the quaternary era, and most notably in the famous description of Axel's dream in Chapter Thirty-Two of that novel. Here, Cuvier is evoked in much the same spirit as Balzac's quasi-poetic figure who resuscitated long-vanished species through an almost magical act of creativity. Axel initially laments his own inability to do the same, but clearly, Cuvier provides the inspiration and the trigger for the dream that follows:

> Je regarde dans les airs. Pourquoi quelques-uns de ces oiseaux reconstruits par l'immortel Cuvier ne battraient-ils pas de leurs ailes ces lourdes couches atmosphériques? Les poissons leur fourniraient une suffisante nourriture. J'observe l'espace, mais les airs sont inhabités comme les rivages. Cependant mon imagination m'emporte dans les merveilleuses hypothèses de la paléontologie. Je rêve tout éveillé. Je crois voir à la surface des eaux ces énormes chersites, ces tortues antédiluviennes, semblables à des îlots flottants. Sur les grèves assombries passent les grands mammifères des premiers jours, le leptotherium, trouvé dans les cavernes du Brésil, le mericotherium, venu des régions glacées de la Sibérie [...].[18]

There is in these lines, and throughout the famous dream sequence that follows, a near-poetic intoxication, and a verbal intensity which fully parallels the Balzacian "éloge de Cuvier" quoted earlier. The passage is certainly one of the most self-consciously poetic in any of Verne's novels, with its conspicuous enumerations, its focus on the sounds of those difficult and strange names, and its cumulative style which, like Balzac's, increases in momentum and speed as it progresses. And as for similarities with Flaubert's more direct verbal response to nature, there are many moments in *Vingt mille lieues sous*

[17] Butor, "Le point suprême", pp. 130-62.
[18] *Voyage au centre de la terre,* p. 274.

les mers in which Aronnax's observation of the flora and fauna, the patterns, colours, sumptuous effects of light and infinite variety of the ocean depths inspire an ecstatic – if not mystical – response in the first-person narrator. Nature itself becomes a source of almost infinite poetry, both because of its sheer abundance, and because of the intrinsic beauty of the spectacle it offers. When Aronnax, in diving gear, accompanies Nemo on a walk on the ocean bed, we have just such a passage:

> Il était alors dix heures du matin. Les rayons du soleil frappaient la surface des flots sous un angle assez oblique, et au contact de leur lumière décomposée par la réfraction comme à travers un prisme, fleurs, rochers, plantules, coquillages, polypes, se nuançaient sur leurs bords des sept couleurs du spectre solaire. C'était une merveille, une fête des yeux, que cet enchevêtrement de tons colorés, une véritable kaléidoscopie de vert, de jaune, d'orange, de violet, d'indigo, de bleu, en un mot, toute la palette d'un coloriste enragé![19]

The reference in the final words here to the "coloriste" makes explicit Verne's wish to convey the spectacle as art, and there is an element of deliberate transposition, here as elsewhere, of natural scene into textual artefact. And while such passages in Verne may not herald an overarching or explicit artistic credo, as they appear to do in the early Flaubert, we may nonetheless sense that they are crucial both to his art and to his vision of the natural world, indeed that they play a fundamental role in eliciting a nuanced and complex response from the reader. Crucially, though, even in these passages that evoke the spectacle of nature in such effusive, essentially poetic terms, there is a gesture in the direction of a more scientific, specialist vocabulary ("plantules, coquillages, polypes"). This is the kind of passage in Verne that so inspired writers like Perec, who once wrote: "Quand, dans *Vingt mille lieues sous les mers*, Jules Verne énumère sur quatre pages tous les noms de poissons, j'ai le sentiment de lire un poème".[20]

Despite and indeed because of the differences between Verne's style and the apparently more "poetic", more "canonical" approach of writers like Balzac and Flaubert, we find in the *Voyages extraordinaires* a new form of discourse, or new discourses, for the novel. Such discourses owe almost everything to the discoveries of travellers, explorers and scientists in his own century and to the new

[19] *Vingt mille lieues sous les mers*, p. 173.
[20] Georges Perec, "J'ai fait imploser le roman", *Galerie des arts*, 184 (1978), 73.

languages that they have initiated. Necessarily messy, extensive, expansive, Verne's style goes in the direction of proliferation rather than or synthesis and unity. It assimilates alien terms, often with an arbitrariness that verges on the cavalier. But herein, precisely, lies the interest and the innovative quality of Verne's work. He is quintessentially of his time in terms of his subject matter, his geographical range, his capturing of the spirit of the age in his stories of exploration and discovery. But he is of his time, too, in terms of his experimental approach to the novel, and in his extension of the very vocabulary and idiom of fiction into the domain of science and knowledge. It is true that, in pursuing this goal, Verne comes close at times to destroying the foundations of fictional coherence and credibility. Yet it is precisely there, in that radical *mise en cause* of what literature now is, or can be, that he most deserves to be taken seriously. There is, in Verne's scientific discourses, an intensely modernistic and sceptical attitude towards the very business of writing. Writing can no longer be "innocent", and through his own "histoires de la terre" Verne is intensely, abundantly, promiscuously and exultantly intertextual. With Verne, it seems, geography and its many related disciplines have definitively invaded the novel and transformed its status.

Jules Verne's Transylvania: Cartographic Omissions

Anca Mitroi

Abstract: *Focussing particularly on Verne's* Le Château des Carpathes *of 1893, this chapter looks more closely at certain geo-political particularities of the author's work. It examines how the depiction of Eastern European territories in Verne's writing may be understood not only in relation to the pervasive orientalist discourses of his time, but also in respect of the author's sentiments regarding (French) nationalism as well as political and cultural imperialism in a wider European context.*

The fact that Verne included Eastern Europe among the strange destinations of the *Voyages extraordinaires* is not too unusual for his time. Théophile Gautier's ironic comments about nineteenth-century rural France being a foreign place[1] make the idea of considering Romania, Hungary or Bulgaria as exotic more plausible. Among Eastern European lands, Transylvania was for many reasons a quasi-mythical[2] place for many writers. Verne refers to it in several novels; in *Le Château des Carpathes* he describes the faraway region, its strange inhabitants and its history in elaborate, almost realistic detail. With a few

[1] See Théophile Gautier, *Voyage pittoresque en Algérie (1845)*, ed. Madeleine Cottin (Genève-Paris: Droz, 1973), pp. 168-69. See also my article, "Fantômes peints et Turcs réels. Illusions perdues d'un voyageur idéaliste", *Bulletin de la Société Théophile Gautier*, 23 (2001), 335-45.

[2] One may say that if in Jarry's *Ubu Roi* Poland means "nowhere", in Verne's times, Transylvania had almost the same status, completely dissolved in the Austro-Hungarian Empire. Moreover, there were no "Transylvanians" since the land was inhabited by a Romanian majority that had no recognized official status and by three other "recognized nations": Hungarians, Saxons and Szekels.

exceptions, the mountains, the green plateaux and the mining towns can be easily recognized and their depictions rival those of geography books.

This attention to detail may have contributed to the general acclaim of the novel in Romania because, while it is true that the fictitious village of Werst – the place where most of the action supposedly takes place – could not be located on any maps, the village's fictionality did not diminish the enthusiasm it inspired in Romanian readers. In her penetrating article "Jules Verne in Transylvania", Roxana Verona examines the reception of the novel, showing that most of the Romanian readers "constantly praised Verne's 'accurate' depiction of the 'real' region, and they have done so since the novel's first translation in 1897".[3] They were obviously able to overlook derogatory comments and, "when Romanian readers choose unconditionally to admire *The Castle,* they implicitly agree to see their status as Europe's 'other'".[4] Apparently, most of the Romanian readers were willing to embrace a book that, in spite of its title, ended up erasing "the real Transylvania's cultural traditions as a province that, at the end of the nineteenth century, was part of the Austro-Hungarian Empire, and as such had rich contacts with Central and Western Europe".[5] If we question what could motivate this reception, we may see that one of the reasons is that, paradoxically, *Le Château des Carpathes* is less about Transylvania than it appears. Verne's geographical omissions and inaccuracies allowed the Romanian readers to reconstruct a political puzzle that was not in Verne's intent, but corresponded to their fin-de-siècle political views.

The assertion that *Le Château des Carpathes* is not really about Transylvania may surprise: the positive Romanian reaction seems motivated precisely by the fact that readers were able to recognize their land and their customs to such an extent that they could ignore the critical comments about the country's economic or cultural backwardness. They could decode the landmarks in the realistic details borrowed – or copied – from Elisée Reclus and Auguste de Gérando, and translate them into existing places, historical sites and

[3] Roxana M. Verona, "Jules Verne in Transylvania", *The Comparatist,* 28 (2004), 135-50 (p. 136).
[4] Verona, "Jules Verne in Transylvania", p. 147.
[5] Verona, "Jules Verne in Transylvania", p. 146.

ethnographic information. Some translators and editors even thought that the fictional information on Transylvania was so flattering and accurate that it needed just a few corrections to be perfect: in order to make every detail better correspond to the Romanian realities, they re-constituted what they thought was the right spelling of towns or rivers, rewrote the idiomatic expressions, and re-baptized the characters (Miriota became Miorita, Gortz became Gorj, Frik, Frig, and so on).[6]

Happy to correct these minor "mistakes", Romanian readers seemed ready to overlook the criticism. As Verona points out, one can still find numerous "negative comparisons [which], though comical, widen the cultural gap between the French knowledge and Romanian ignorance, as one country obviously 'has' what the other 'lacks'"[7] as the novel seems to expose the "lack of vegetation, lack of civilization, of humanity, and of knowledge".[8]

However, even if Verne's portrayal of Transylvania does not always seem very flattering (for instance, Werst is designated as "l'un des plus arriérés villages du comitat de Koloswar",[9] untouched by Western civilization), we may say that such images are still more positive than many other of Verne's "extraordinary" destinations. As Jean Chesneaux points out, Verne often treats foreign nations or ethnic groups with an openly racist disdain based, probably, on the belief of the superiority of Western Europe,[10] or, more specifically, that of

---

[6] Verona, "Jules Verne in Transylvania", pp. 141-42; Verona examines various translations and interpretations, showing that: "When Hobana, the knowledgeable specialist of Verne's works, reviews Romanians' animated reception of the novel, he notices that some translators tried to give it a Romanian turn by changing the Hungarian or German names into Romanian ones and correcting any topographical errors. While Verne is Gothicizing the local, Romanian translators are localizing the Gothic" (p. 145). See also Raluca Anamaria Vida, "*L'Île mystérieuse – Insula misterioasa* comme paradigme du phénomène retraductif roumain dans le cas de Jules Verne", *Jules Verne dans les Carpates. Caietele de l'Echinox.*, vol. 9 (Cluj: Universitatea Babes-Bolyai, 2005), pp. 262-72, and Muguras Constantinescu, "Remarques sur la traduction en roumain du *Château des Carpathes* de Jules Verne", *Atelier de traduction*, 3 (2005), 99-111.

[7] Verona, "Jules Verne in Transylvania", p. 142.

[8] Verona, "Jules Verne in Transylvania", p. 139.

[9] Verne, *Château*, p. 8.

[10] Jean Chesneaux talks about "un mépris quasi raciste" (p. 18) and of "tout l'Occident du machinisme moderne et du capitalisme ascendant qui crie son orgueil d'être 'européen' et qui étend son contrôle à la surface de la planete – ou du moins croit naïvement qu'il pourra le faire sans encombres". (p. 22). "*Le Tour du monde en*

France as a civilizing, colonial power. In *Le Château des Carpathes*, the two Romanian protagonists, Franz de Telek and Rodolphe de Gortz, have noble traits, Western education and even a certain degree of sophistication. This is not only the case for the upper classes: with some exceptions – like Frik, the shepherd, depicted more like a beast than a human, hairier than his own sheepskin cloak, and "aussi mal tenu de sa personne que ses bêtes"[11] – the representatives of the lower class can also be endowed with undeniably positive qualities: both Miriota, in her simple yet seductive peasant attire and her fiancé, Nic Deck, are presented as prototypical Romanians and depicted as attractive, honest and intelligent, although a bit naïve and superstitious, like anybody in rural Romania, as the narrator implies.

The vaguely Westernized characters who are still able to return to ancestral customs and beliefs once they go back to their fatherland, together with characters like Miriota or Nic, who cling to their faith and traditions in the midst of all the turmoil, make Verne's attitude ambiguous at the very least: the Romanians are presented as rather primitive indeed, and yet this primitivism may have been another element that contributed to the novel's positive reception. In the portrayal of the Romanians' unadulterated ways, the readers could perceive a nostalgic tone: the setting of *Le Château des Carpathes* is placed beyond time, outside of history, and Verne thus creates a "paradigme du site hors espace et voué à une temporalité a-historique".[12] Paradoxically, what was described in such minute detail turns out to be an abstraction, or an indirect way of nostalgically evoking the pre-modern qualities of an idealized Ancien Régime France. Verne invests a few imaginary or foreign lands with the qualities of a romanticized old France; Joëlle Dusseau remarked this perspective in other Vernian novels in which the narrator seems to describe foreign lands (like Canada) when he is actually talking about his homeland: "Le Canada c'est la France, mais c'est la France de l'Ancien Régime."[13] This is why, as Verne talks about the long lost traditions of the Western world, or about preserving various foreign customs un-

---

quatre-vingt jours Notes de lecture", *La revue des lettres modernes, Jules Verne. Le Tour du monde. La revue des lettres modernes,* (3) nos 456-61 (1976), 11-20.

[11] Verne, *Château*, p. 9.

[12] Jean-Pierre Picot, "*Le Château des Carpathes*: influences, confluences, effluences", *Caietele.Echinox.* Vol. 9, *Jules Verne dans les Carpates*, p. 37.

[13] Joëlle Dusseau, *Jules Verne* (Paris: Perrin, 2005), p. 301.

changed for centuries, we may detect an allegorical critique of modernity's displacement of Ancien Régime traditions. If Frik, the almost savage shepherd, could be a "descendant dégénéré" of his noble Dacian ancestry, the river by which his sheep are grazing is nevertheless still so pure that the narrator easily imagines it "couler à travers les meandres du roman de *l'Astrée*".[14] Verne's reference to the *dolmens* and *menhirs* of Carnac may also allude nostalgically to the same idea of a land that still preserves the purity of an archaic, long-gone France. At the same time, for many Romanian readers, such descriptions of a world unchanged for centuries were not necessarily seen as a critique. In the nineteenth century, many nationalist discourses focused on the *continuity* of the Romanians in Moldavia, Walachia and Transylvania, an idea that was often based on the preservation of unadulterated Dacian traditions and costumes by the peasants of the three territories.

The Transylvania of *Le Château des Carpathes* thus differs not only from other foreign lands derogatorily depicted in Verne's novels, but also from many images of Eastern Europe described by other authors around that time and in which Romanian readers refused to recognize themselves. As Larry Wolff argues in his *Inventing Eastern Europe,* since the eighteenth century, most of the descriptions of anything east of Austria abounded with grotesque, quasi-fantastic imagery meant to underscore the difference with the "civilized world" of Western Europe, and to prove that "Eastern Europe was a realm of fantastic adventures with savage beasts, whose wildness [could be] triumphantly tamed in a parable of conquest and civilization".[15] That was the case of Walachia and Moldavia which, even in the nineteenth century, continued to inspire images of a fictional Oriental world invented by eighteenth-century literati.

Because of the persistence of the idea of an exotic Eastern Europe in the nineteenth century, many journals of Western travellers simply included the two Romanian principalities – Walachia and Moldavia – in the Ottoman Empire, and sometimes continued this "Orientalization" even after Romania's Independence War. In 1878, a Romanian diplomat underscored this anachronistic view: "Jusqu'ici, l'Angleterre ne s'est intéressée à la Roumanie que sous le point de vue

[14] Verne, *Château*, p. 8.
[15] Larry Wolff, *Inventing Eastern Europe, The Map of Civilization on the Mind of the Enlightenment* (Stanford: Stanford University Press, 1994), p. 101.

de partie intégrante de l'Empire ottoman, et non point comme à un Etat roumain, à un peuple qui a des bonnes et utiles aspirations."[16] In order to oppose this inclusion in the Orient, he added: "La Roumanie a prouvé à l'Europe qu'elle veut et qu'elle peut marcher dans la voie du progrès et de la civilisation occidentale [....] Les Roumains sont de race latine."[17] However, the clichés of an "Oriental Romania" were too picturesque to be easily effaced: Raoul Perrin's 1839 *Coup d'œil sur la Valachie et la Moldavie*, for example, portrayed Bucharest with countless Turkish baths managed by turbaned genies identical to those known to French readers from Antoine Galland's translation of the *One Thousand and One Nights*.[18] According to the same traveller, when not greeting each other the Oriental way, with a "Sélam alékoum", the inhabitants of these countries either spoke a mix of Slavic, Italian, Latin and barbaric words,[19] or just made various fearsome growls and snarls,[20] a view which differs significantly from Verne's insistence on the Latin origin of the Romanian language. According to Perrin, the Romanians' clothing was described as an exotic mix with no individual identity: "the Walachian worker may wear a Turkish turban, a Greek fur cap, Armenian sandals, a Bulgarian belt, a Crimean coat and Albanian pants, so that for the Europeans, this bizarre confusion provides a show more attractive than our cheap carnivals",[21] while for Verne, the costumes worn by Miriota or judge Koltz are perfect definitions of their Romanian nationality. For French readers seeking hair-raising thrills, Perrin and other travellers provided more exotic accounts of the bestial tortures being inflicted by the same carnival-like turbaned Romanians: cut eyelids, clothes sewn to the sufferer's skin, macabre bowling games with the victims' heads and, of course, thousands and thousands of deaths by impaling.[22]

If none of these gory scenes or exaggerated exotic costumes appear in Verne's Carpathian village, it is not necessarily because the Western opinion of Romania had dramatically changed by Verne's

---

[16] Mihail Kogalniceanu, *Acte si documente din corespondenta diplomatica a lui Mihail Kogalniceanu relative la rasboiul independentei Romaniei 1877-1878*, (Bucharest: M. Kogălniceanu, 1893), p. 47.

[17] Kogalniceanu, *Acte si documente*, p. 44.

[18] Raoul Perrin, *Coup d'oeil sur la Valachie et la Moldavie* (Paris: A. Dupont, 1839).

[19] Perrin, *Coup d'œil*, p. 5.

[20] Perrin, *Coup d'œil*, p. 15.

[21] Perrin, *Coup d'œil*, p. 24.

[22] Perrin, *Coup d'œil*, p. 28.

time. The images recalling *One Thousand and One Nights* were successful enough that several years after Perrin's book, other authors, such as Stanislas Bellanger, copied the descriptions word for word, adding, nevertheless, that they had been eyewitnesses to the exotic horrors plagiarized from the *Coup d'œil sur la Valachie et la Moldavie*.[23]

Verne's perspective is thus similar to that of a few authors who, instead of insisting on the frightening and savage side of Eastern Europe, invented something that was more of a recollection of a pure (yet primitive and also fictitious) ancient land. An example could be Princess Soubiran-Ghika's books on Romania. In the opening pages of *La Valachie devant l'Europe*, as she witnesses a religious procession, she meditates on a country that, through its archaic customs, is still "une civilisation enfantée par l'idée chrétienne".[24] In her eyes, Walachia is, quite like Verne's Transylvania, "[un] pays de contrastes qui a le privilège bien rare aujourd'hui d'avoir conservé intacte sa physionomie particulière, au milieu même du développement de la civilisation",[25] and which still preserves ancient practices like "cette pratique touchante de consacrer au service de Dieu la dîme de sa fortune, trop perdue en Occident, [qui] vient ici rappeler les premiers âges du christianisme et l'ère de la foi".[26] The same archaic Christianity reminiscent of Nerval's *Sylvie* is to be found in Verne's Carpathians where, for instance, the young girl Miriota, when afraid, makes a sign of the cross, "se sign[ant] du pouce, de l'index et du médius, suivant cette coutume roumaine, qui est un hommage à la Sainte Trinité".[27]

Soubiran-Ghika thought that the Romanians had preserved such ancient values and found in nineteenth-century Romania a pre-Revolutionary version of France. Her journey was thus more like a journey in time since in the Romanian sites she constantly sees a long lost fatherland resonating with the fairytales of her childhood.[28] The idea of fairytales is connected with the past, and the regrettable disap-

---

[23] See Stanislas Bellanger, *Le Keroutza, voyage en Moldo-Valachie* (Paris: Librairie Française et Étrangère, 1846).

[24] Aurélie Soubiran-Ghika, *La Valachie devant l'Europe* (Paris: Dentu, 1858), p. 8.

[25] Aurélie Soubiran-Ghika, *La Valachie moderne* (Paris: Comon, 1850), pp. 25-26.

[26] Soubiran-Ghika, *La Valachie moderne*, p. 226.

[27] Verne, *Château*, p. 62.

[28] Verne, *Château*, p. 243.

pearance of this tradition is seen as a loss characteristic of the modern age. Opposing the oral tradition of fantastic stories in Romania and modern France, Ghika comments: "On ne conte plus. Quelques vieillards ont conservé ce talent comme le souvenir d'un siècle éteint. On ne cause plus, on pérore: le forum a tout envahi. Nous mourrons par la parole."[29] Verne's commentary on traditional storytelling is quite similar, as he nostalgically deplores the disappearance of legends in a too pragmatic nineteenth century: "D'ailleurs, il ne se crée plus de légendes au déclin de ce pratique et positif XIXe siècle [...]".[30] However, if the villagers of Werst do not invent new legends, Verne explains that they preserve and cultivate their traditions and "ces légendes qui prennent volontiers naissance dans les imaginations roumaines".[31] Paradoxically, such considerations about the naïve attachment to legends and even the lack of sophisticated education might have been seen positively by Romanian readers: in nineteenth-century Transylvania, as "[t]rue Romantics, more and more Romanians turned to the unspoiled 'people' as the repository of some pristine national wisdom and virtue".[32] If the Transylvania of the *Château des Carpathes* seems to be inhabited by "noble savages" as relics of a frozen past, this should not surprise us: Jean-Pierre Picot points out that Verne often sees in certain territories, like Scotland for Paganel, or in this case, Transylvania, "[un] haut lieu du passé".[33] This also means that even if Verne totally effaces the real Transylvania, its actual economic and cultural development and its ties with Western Europe, he creates an imaginary land similar, on one hand, to Nerval's idyllic Valois with all its legends and relics of the past, and, on the other, to the Romantic discourses of Dacian legends and illusory pure origins which are so familiar to Romanian readers.

Yet the most striking omission in Verne's Transylvania is one that could be taken either for an inexplicable mistake or for a political statement, that is the total absence of Hungarians and Austrians in a

---

[29] Soubiran-Ghika, *La Valachie moderne*, p. 19.
[30] Verne, *Château,* p. 8.
[31] Verne, *Château,* p. 29.
[32] Gabro Barta and Istvan Bona, *History of Transylvania,* trans. by Adrienne Chambers-Makkai et al. (Budapest: Académiai Kiadó, 1994), p. 473. On this aspect of Romanian nationalism, see also Anne-Marie Thiesse, *La Création des identités nationales: Europe XVIIIe-XXe siècle* (Paris: Seuil, 1999), pp. 95-100.
[33] Jean-Pierre Picot, "Parodie et tragédie de la regression dans quelques œuvres de Jules Verne", *Romantisme. Revue du Dix-neuvième siècle,* 27 (1980), 109-27 (p. 111).

land that, at the time, was a part of the Austro-Hungarian Empire. From a modern (or anachronistic) perspective we could see how, in their enthusiasm, the Romanian readers embraced this omission from a nationalist and patriotic point of view. To twentieth-century readers, it may appear natural to talk about Romania in relation to *Le Château des Carpathes*. But to Verne's contemporaries, this should have been seen at least as an anachronism, since Transylvania had become a part of Romania only in 1918. Subsequently, its northern part was taken back by Hungary in 1940, and was returned to Romania only after the Second World War.

The national identities in *Le Château des Carpathes* are thus perplexing: "While the local population speaks Romanian, as Verne clearly indicates, most of the place names are Hungarian (Koloswar, Maros, Thorda) and most of the characters' names are German (Werst, Frik, Koltz, Deck, Gortz, Franz)."[34] We may think that he is confused about the Romanian names since in *Le Beau Danube Jaune* one of the characters is called "le Hongrois Miclesco": Miclesco is the French spelling of a very common Romanian name. At the same time, how confused or misinformed could Verne be since he is the author who seemed to be informed about everything? Roxana Verona raises this question: "Do such errors indicate a lax attitude toward the accuracy of sources, as Hobana claims? Is this negligence toward local names the sign of an ideological stance – an indifference toward the places of reference – or is it simply that Verne sped through the process of writing without double-checking his facts?"[35]

In 1895, one of Verne's contemporaries, Charles Canivet, commented on the writer's excellent documentation and concern with world politics: "Qu'un fait plus ou moins important se produise en un point quelconque du globe, ou bien en quelque endroit isolé des plus vastes mers, il n'est pas douteux que Jules Verne ait passé par là."[36] As an author whose fiction rivals with geography or history books, Verne may appear here almost like his own character, Paganel, who inadvertently learned Portuguese instead of French: his documentation seems accurate, yet his version of Transylvania, as a land inhabited only by Romanians, diverges from historical fact. Not only do all the

---

[34] Verona, "Jules Verne in Transylvania", p. 141.
[35] Verona, "Jules Verne in Transylvania", p. 146.
[36] Jean-Michel Margot, *Jules Verne en son temps vu par ses contemporains francophones (1863-1905). Cahiers Jules Verne II* (Amiens: Encrages, 2004), p. 194.

characters speak Romanian, as Verona points out, but mostly everything they refer to in their daily lives is Romanian. Even the mythical creatures they believe in ("babe", "staffii", "serpi de casa" or "balauris"[37]) have Romanian names. Franz de Telek, in spite of his German / Hungarian name has his roots in Krajowa (Craiova), which is not in Transylvania but in Walachia: "Krajowa est une des principales bourgades de l'Etat de Roumanie, qui confine aux provinces transylvaines vers le sud de la chaîne des Carpathes. Franz de Telek était donc de race roumaine – ce que Jonas avait reconnu au premier aspect."[38]

Moreover, the legends evoked in the novel, such as those concerning "Miorita" and "Manole" are not just from the Romanian Kingdom, but were seen by Romanians as founding myths of Romanian identity. As Anne-Marie Thiesse explains in *La Création des identités nationales,* such legends were written or "discovered" by Romanian Romantic nationalists like Vasile Alecsandri and Alecu Russo at a time when many European nations were patriotically "inventing" their national past. Romantic poet and revolutionary politician, Vasile Alecsandri, whose biography bears an uncanny resemblance to that of Rodolphe de Gortz, not only tried to establish Romania's mythical origins through these legends, but was also active in developing cultural and political ties with France, by writing the famous song of "The Latin Race" (Cântecul gintei latine).[39] By attributing the construction of the Carpathian Castle to the legendary architect "Manoli", Verne establishes a cultural and artistic heritage common to Walachia and Transylvania, which could only please Romanian readers: "Quel architecte l'a édifié sur ce plateau, à cette hauteur? On l'ignore, et cet audacieux artiste est inconnu, à moins que ce soit le Roumain Manoli, si glorieusement chanté dans les légendes valaques, et qui bâtit à Curté d'Argis le célèbre château de Rodolphe le Noir."[40]

---

[37] Verne, *Château*, pp. 28, 103.
[38] Verne, *Château*, p. 111. We may add here that through Telek's name, Verne is alluding to his source of documentation for his novel, Gérando, who was married to a Hungarian aristocrat from the Teleki family.
[39] See Anne-Marie Thiesse's comments on creating the Romanians' nationality as "sons of the Dacians" and Vasile Alecsandri in the revolutionary moment of 1848, in *La Création des identités nationales*, pp. 95-100.
[40] Thiesse, *La Création des identités nationales*, pp. 25-26.

Although Jean-Paul Dekiss quite strangely claims, in spite of all evidence, that "Le comte Franz de Telek, Mathias Sandorf et le baron de Gortz sont tous trois d'origine hongroise",[41] the main characters are not just Romanians, but great patriots. Judge Koltz's house is in itself a nationalist political statement, as its walls are decorated with the "portraits violemment enluminés des patriotes roumains".[42] Franz de Telek is a "descendant d'une noble famille de race roumaine",[43] that is from "l'une des plus anciennes et des plus illustres de la Roumanie".[44] Rodolphe de Gortz is equally a great defender of the Romanians of Transylvania:

> Les barons de Gortz étaient seigneurs du pays depuis un temps immemorial. [...] Ils luttèrent contre les Hongrois, les Saxons, les Szeklers; ils avaient pour devise le fameux proverbe valaque: *Da pe maorte*, "donne jusqu'à la mort!" et ils donnèrent, ils répandirent leur sang qui leur venait des Roumains, leur ancêtres.[45]

After he receives a Western education, the noble Romanian returns to his native country to support the anti-Hungarian resistance: "Il n'avait pas oublié la patrie transylvaine au cours de ses lointaines pérégrinations. Aussi revint-il prendre part à l'une des sanglantes révoltes des paysans roumains contre l'oppression hongroise."[46] Although the allegedly famous saying is approximately quoted and misspelled, the proverb resonates with other political considerations that seem to lead to the conclusion that, in the intricate and perpetually unresolved ethnic conflicts in Transylvania, Verne supports the Romanian cause. Could this be possible, given the fact that the novelist otherwise avoided political comment in his works? Romanian readers thought so, and some were ready to imagine a secret affective tie that linked Verne to Romania.

Jean Chesneaux explains: "'Les Transylvains', dit Verne, n'ont plus d'existence politique. Trois talons les ont écrasés.' [...] Trois talons...[soient] vraisemblablement les Romains, les Turcs, les

[41] Jean-Paul Dekiss, *Jules Verne enchanteur* (Paris: Editions du Félin, 1999), p. 292.
[42] Dekiss, *Jules Verne enchanteur*, p. 40.
[43] Verne, *Château*, p. 142.
[44] Verne, *Château*, p 122.
[45] Verne, *Château*, p. 26.
[46] Verne, *Château*, p. 27.

Habsbourg."[47] However, it is unlikely that Verne sees the Romans as enemies of the Romanians, who actually used their double origin (Dacian and Roman) in their revolutionary political discourses. Here is Verne's first presentation of a Romanian Transylvania:

> Tel est cet ancien pays des Daces, conquis par Trajan au premier siècle de l'ère chrétienne. L'indépendance dont il jouissait sous Jean Zapoly et ses successeurs jusqu'en 1699, prit fin en 1699 avec Léopold Ier, qui l'annexa à l'Autriche. Mais, quelle qu'ait été sa constitution politique, il est resté le commun habitat de diverses races qui s'y coudoient sans se fusionner, les Valaques ou Roumains, les Hongrois, les Tsiganes, les Szeklers d'origine moldave, et aussi les Saxons que le temps et les circonstances finiront par "magyariser" au profit de l'unité transylvaine.[48]

In this description, Verne ironically refers to Hungary as a "colonizing" power, and touches a very sensitive topic in the nationalist discourse of the Romanians in Transylvania, that is the use of the Hungarian language as a colonizing instrument. In the *History of Transylvania,* this process is presented in the following terms:

> The attempt to give Hungarian wider currency as the "national" language was an affront to the Romanians' own hopes of national self-realization. The leading Hungarian Liberals attempted in vain to dissociate themselves from any attempt to force Magyarization. The Romanians could only see the proposed measures as the first steps in that direction [...].[49]

The Romanians' resistance to Magyarization could then find its echoes in *Le Château des Carpathes*, especially when Verne foresees a great future for the Romanian nation:

> On le sait, tant d'efforts, de dévouement, de sacrifices, n'ont abouti qu'à réduire à la plus indigne oppression les descendants de cette vaillante race. [...] Mais ils ne désespèrent pas de secouer le joug, ces Valaques de la Transylvanie. L'avenir leur appartient, et c'est avec une confiance inébranlable qu'ils répètent ces mots, dans lesquels se concentrent toutes leurs aspirations: *Rôman on péré!* "Le Roumain ne saurait périr!".[50]

---

[47] Jean Chesneaux, "Microcosme et macrocosme: le statut du *Château des Carpathes* dans la vision du monde et de la société", *Caietele Echinox, Jules Verne*, vol. 9, p. 22.
[48] Verne, *Château*, p. 20.
[49] Barta and Bona, *History of Transylvania,* p. 475.
[50] Verne, *Château*, p. 26.

Of course, patriotic Romanian readers could read this message (although misspelled again) as sympathetic encouragement, especially in a time when: "With Romanian independence achieved there was an upsurge in romantic nationalist feeling beyond the Carpathians. So-called Daco-Romanian calendar maps became increasingly popular, showing the Romanian population between the Black Sea and the Tisza river as one continuous block."[51] When *Le Château des Carpathes* was published in French, and later translated into Romanian, the movement of the Romanians in Transylvania was reaching its peak with the "Memorandum Movement" which issued a petition addressed to Franz Joseph, published in several languages, voicing the Romanians' grievances and receiving considerable international attention. This Memorandum came as the conclusion to a century or so of requests, conflicts, and revolts, and thus, when Verne announces that the future belongs to the Romanians, when in his Transylvania there are no Hungarians and no Austrians, when his linguistic or ethnographic documentation sees no borders between the Romanian Kingdom and Transylvania, his words are perceived by the Romanian readers as a positive political statement, on issues that are still debated even nowadays.

However, we can say that Verne is neither pro-Romanian, nor against the Austro-Hungarian Empire. If we examine other Vernian texts we can see other deliberate cartographic omissions that seem to contradict that of *Le Château*: if the Hungarians and Austrians are absent from his Transylvania, Joëlle Dusseau notices a similar absence in *Le Secret de Wilhem Storitz*: "Là, encore, c'est le sentiment anti-allemand qui domine, même si Verne différencie parfois les Allemands des Autrichiens, qui sont ici les grands absents (de façon bien étonnante dans le royaume austro-hongrois…)."[52] Dusseau underlines other political peculiarities in Verne, thus showing that the absence of Hungarians in *Le Château* does not carry any anti-Hungarian / pro-Romanian meaning: since "Les Magyars de l'Empire d'Autriche sont l'objet d'une autre forte sympathie, dans deux romans, *Mathias Sandorf* et *Le Secret de Wilhem Storitz*."[53] But there is also an exaltation of Greek or Bulgarian nationalism.[54] Therefore,

[51] Barta and Bona, *History of Transylvania,* p. 617.
[52] Dusseau, *Jules Verne*, p. 292.
[53] Dusseau, *Jules Verne*, p. 292.
[54] Dusseau, *Jules Verne*, p. 295.

Verne seems to favour the Hungarians in *Storitz,* yet excludes them from his fictional Transylvania just as he did with the Austrians in *Wilhem Storitz*. The contradiction can be solved if, in these novels, we see a message inspired more from "la solide fidélité 'quarante-huitarde' dont Verne a maintes fois témoigné envers les luttes des peuples pour leur liberté".[55] Dusseau concludes that Verne's political attitude has more to do with France than with the countries that he describes in his *Voyages extraordinaires*:

> Le printemps des peuples, dont Verne a été contemporain, n'est représenté ni dans sa dimension sociale (abolition du servage) non plus que démocratique (constitutions, élections), ni dans son jaillissement d'aspirations identitaires multiculturelles. Il se réduit à une attitude exclusivement anti-allemande, expression non d'une réalité mais du sentiment qui se développe en France et que tout au moins Jules Verne porte en lui. Un sentiment tardif, surtout présent à partir de 1885, bien des années après la défaite, la perte de l'Alsace-Lorraine, l'occupation, précisément au moment où le boulangisme se développe. Le mouvement des nationalités, bien présent chez Verne, est essentiellement vu à travers le prisme de l'hostilité au vainqueur de 1870.[56]

However, we may add that Verne is, especially in his late years, rather sceptical or even bitter about nationalist discourses, whatever their origin. In *Le Beau Danube jaune*, the narrator ironically describes the conflicts triggered by frivolous arguments between initially peaceful fishermen whose aggressive behaviour is suddenly ignited by nationalist feelings:

> De plus, les Moldaves de la Société ayant pris fait et cause pour le Moldave, et les Serbes pour le Serbe, il s'en suivit une regrettable bataille qui ne fut pas réprimée sans peine. Il est vrai, de la part de ces pêcheurs à la ligne, qui passent pour des gens si calmes, si placides, si en dehors des violences humaines, tout est possible quand leur amour-propre est en jeu![57]

Founded with the best intentions of international understanding, the peaceful fishermen's society is devoured by personal interests, pride,

---

[55] Jean Chesneaux, "Microcosme et macrocosme", p. 22. If Verne is critical of British colonialism, he is however biased when talking about French colonialism. On this aspect, as well as on Verne's political attitudes, see Timothy Unwin, *Jules Verne, Journeys in Writing* (Liverpool: Liverpool University Press, 2005), p. 21.
[56] Dusseau, *Jules Verne*, p. 299.
[57] Jules Verne, *Le Beau Danube jaune* (Paris: Archipel, 1997), p. 34.

and intolerance, and could be seen as a Vernian meditation on Europe's incapacity to overcome its old nationalisms and destructive enmities. In this sense, it is paradoxical that *Le Château des Carpathes* was received in Romania with great nationalist pride, since the novel may actually carry a rather bitter message that has little to do with Romania and more with a frail European continent. In *Le Château des Carpathes* could Verne be alluding, through Franz and Rodolphe, to the Emperor Franz Joseph and to his son, Rudolph of Austria, who was the Crown Prince? Could he be referring allegorically to their alleged rivalry (since it is said that Franz Joseph's adversaries wanted Rudolph to replace his father at the Austro-Hungarian throne) that had ended in a still unsolved murder / suicide mystery – the famous Mayerling scandal of 1889, when Rudolph was found dead in his hunting lodge? Did Rudolph commit suicide together with his mistress or was he killed for political reasons since he was supporting the idea of an alliance with France, while his father, Franz-Joseph favoured Germany?[58] It is hard to say if *Le Château des Carpathes* reflects any of those contemporary events that had shaken Europe. In any case, as the Carpathian Castle is destroyed, Rodolphe, the only male heir of his family, just like Rudolph of Austria, dies and together with him, an entire world collapses, leaving only ruins and "souvenirs de cet inoubliable passé".[59] Yet, beyond the references to Transylvanian towns or Romanian history and folklore, in Rodolphe and Franz, torn apart by jealousy and selfish desire, it is possible to perceive a tragic image of a self-destructing Europe annihilating itself through endless wars and rivalries.

[58] See Richard Barkeley, *The Road to Mayerling: The Life and Death of Crown Prince Rudolph of Austria* (London: Phoenix Press, 2003).
[59] Verne, *Château*, p. 212.

Undermining Body and Mind? The Impact of the Underground in Nineteenth-Century Children's Literature

Kiera Vaclavik

Abstract: *Considering not only Verne's* Voyage au centre de la terre *but also lesser known works by George MacDonald, Hector Malot, R. M. Ballantyne and Mme de Gériolles, this chapter analyses how the age-old narrative device of a descent into the underworld is employed to popularize science for a young audience. These texts not only reflect contemporary thinking on the geological composition of the earth's crust and core but also play, via their narrative forms, with notions of geological time and the relative age and development of the human species.*

In classical literature, the journey down into the underground, known to the Ancient Greeks as *katabasis*, was also a journey back in time. Virgil's Aeneas, for example, "journeys amidst the sorrows of the past" in his encounters with Palinurus, Deiphobus and Dido.[1] Moreover, the episode can itself be seen as an intertextual return, harking back to the *katabasis* of book XI of the *Odyssey* where Homer's hero also revisits figures from his past.[2] Journeys like those of Odysseus

[1] R. D. Williams, "The Sixth Book of the *Aeneid*", in *Oxford Readings in Vergil's* Aeneid, ed. by S. J. Harrison (Oxford/New York: OUP, 1990), pp. 191-207 (p. 194).

[2] At the beginning of book XI of the *Odyssey*, widely referred to as the *nekyia*, it seems that it is the dead who rise up to Odysseus rather than the hero who journeys down to the underworld. Yet the subterranean situation of the land of the dead is firmly established in the course of the narrative and, even though it is impossible to establish the precise point of departure, it is nevertheless clear from both the Greek

and Aeneas are not restricted to the classical period, however, but continued to resonate and be recounted in the modern age. As several critics have observed, the nineteenth century was a particularly fertile period for the katabatic narrative.[3] Less widely remarked is the fact that *katabasis* was incorporated not only into adult works but also those targeting a family audience or specifically juvenile readership. Indeed some of the best-known, most celebrated works of the period such as Lewis Carroll's *Alice's Adventures in Wonderland* (1865, originally entitled *Alice's Adventures Underground*) and Jules Verne's *Voyage au centre de la terre* (1864) are built around a journey down to and back from an underground locale. In *Voyage au centre de la terre*, as in classical *katabases*, the journey through space is also a journey back in time, although as I have argued elsewhere, the personal dimension is eliminated and the temporal parameters vastly extended.[4]

Children's literature also participates in the development in this period of another form of underground story identified by Rosalind Williams in which the underworld becomes a place to live or work instead of a place to visit.[5] Rather than the extraordinary destination of a once in a lifetime journey, the underground becomes a place where a considerable amount of time is spent.[6] As Williams has

text and English translation that the hero has passed into this underground realm. When asked by his mother, "what *brings you down to* this world of death and darkness?", Odysseus's equally revealing response is that he "had to *venture down to* the House of Death,/ to consult the shade of Tiresias" (my emphasis, Homer, *The Odyssey*, trans. by Robert Fagles (London: Penguin, 2001), XI, pp. 177, 186. Odysseus's presence in the underground is reinforced by the repeated use of the determiners "here" and "this", indicating proximity and immersion rather than distance.

3 Rosalind Williams, *Notes on the Underground: An Essay on Technology, Society, and the Imagination* (Cambridge, MA/London: MIT Press, 1990); Wendy Lesser, *The Life Below the Ground: A Study of the Subterranean in Literature and History* (Boston/London: Faber & Faber, 1987); Lyle Thomas Williams, "Journeys to the Center of the Earth: Descent and Initiation in Selected Science Fiction" (unpublished doctoral dissertation, University of Indiana, 1983); Walter Strauss, *Descent and Return: The Orphic Theme in Modern Literature* (London: OUP, 1971).

4 Kiera Vaclavik, "Jules Verne écrivain… de jeunesse: The Case of *Voyage au centre de la terre*", *Australian Journal of French Studies*, 42 (2005), 276-83.

5 Williams, *Notes on the Underground*, p. 10.

6 The colonization of the underground in these texts had real-life precedents: the American physician John Corghan bought the famous Mammoth Cave in 1839 where he installed consumptives in wooden and stone huts in order that they benefit from the constant temperature and humidity thought to have significant healing properties. See

shown, these underground narratives constitute projections into the future, but the result of subterranean existence is frequently a movement backwards rather than forwards, and indeed life below ground can itself be seen as a return to a primitive condition given that "[h]umanity's earliest constructions were burrows rather than buildings".[7] In literary works, H. G. Wells's Morlocks are perhaps the best example of the underground's generation of regression rather than progression when adopted as permanent dwelling place, while Zola's *Germinal* offers a similar portrayal of the effects of a subterranean working environment.

According to theories of recapitulation which became popular in the late nineteenth century, children were themselves situated at the foot of the evolutionary ladder with a journey up to the summit of fully-formed adulthood before them.[8] Given that one of the traditional functions of children's literature is that of socialization, of assisting the process of transition from childhood to adulthood, the way in which texts for young readers present the impact of the underground environment therefore merits examination. In works for young readers specifically, to what extent does going down below for these substantial periods of time also entail going back – in a physical, moral, evolutionary sense – or is a more positive, progressive vision apparent? This question will be explored with reference to English and French texts by Jules Verne, George MacDonald, Hector Malot, R. M. Ballantyne and Mme de Gériolles dating principally from the 1870s which, in all but one case, take coal mines as their underground locales. Having examined texts which present the underground principally as dwelling place, the focus will then move to those in which the underground constitutes a place of work.

George MacDonald's *The Princess and the Goblin* (1872), relates the story of Princess Irene, her relationship with her magical grandmother and with the young miner, Curdie, who uncovers and is eventually instrumental in foiling a goblin plot to abduct the heroine. There are

John F. Sears, *Sacred Places: American Tourist Attractions in the Nineteenth Century* (Amherst: University of Massachusetts Press, 1998), p. 37.

[7] Williams, *Notes on the Underground*, pp. 16, 90.

[8] Julia Briggs, "Transitions: 1890-1914", in *Children's Literature: An Illustrated History*, ed. by Peter Hunt (Oxford/New York: Oxford University Press, 1995), pp. 167-91 (p. 169).

clear undertones of Persephone and her abduction by Hades in this text, since the goblin population live in the subterranean caverns beneath the princess's castle.[9] The opening chapter of the text is almost entirely devoted to the nature of the goblins and the reasons for their underground existence: according to local legend, the goblins used to live "above ground, and were very like other people" but, in response to what they perceived as unjust treatment by the king, moved below ground, the caverns explicitly serving as a "refuge".[10] However, long-term subterranean existence has had dramatic consequences for both the physical and moral condition of the goblins: "they had greatly changed in the course of generations; and no wonder, seeing they lived away from the sun, in cold and wet and dark places" (p. 4). They have gained strength and, it seems, longevity – a father goblin says to his child at one point: "Your knowledge is not quite universal yet […] You were only fifty last month. Mind you see to the bed and bedding" (p. 55). But as a result of living underground they have also "sunk" towards their animal companions (p. 101). Stunted, with hard heads, tender, toe-less feet, nail-less hands and mole eyes no longer able to tolerate sunlight, they are "not ordinarily ugly, but either absolutely hideous, or ludicrously grotesque both in face and form" (p. 4). Their physical degeneration is accompanied by moral and spiritual disease: they have grown cleverer in the underground, but their knowledge takes the form of "cunning" rather than wisdom (p. 4). Their time is spent devising ways in which to torment the residents of the surface and they revel in the misfortune of others, even that of other goblins ("for a few moments the others continued to express their enjoyment of [the goblin prince's] discomfiture"(p. 135)). They consider themselves to be vastly superior to their opponents on the surface, who, tellingly, regard the goblins as a "degraded" race (p. 67).[11] Focalized

[9] The presence of the abduction narrative in the text is discussed by Joseph Sigman, "The Diamond in the Ashes: A Jungian Reading of the *Princess* Books", in *For the Childlike: George MacDonald's Fantasies For Children*, ed. by Roderick McGillis (Metuchen/London: The Children's Literature Association and The Scarecrow Press, 1992), pp. 183-94.

[10] George MacDonald, *The Princess and the Goblin* (London: Puffin, 1996), pp. 2, 4. Subsequent references are to this edition and are given after quotations in the main body of the text.

[11] Sigman regards the goblins, with their hard heads, aversion to poetry and sense of superiority as "images of the sceptical and materialistic tendencies in Victorian society" (p. 188). In other words, the goblins are Philistines, and thus, according to this

through Curdie, the young miner who does not fear but rather mocks the goblins, they are frequently absurd and ridiculous rather than terrifying. But they nevertheless remain a tangible and potent threat to the princess and her people.

It is not just the goblins themselves who have been affected by life underground but also their household animals known as the cobs creatures. The ancestors of these creatures were domestic animals or tamed wild animals taken by the ancestors of the goblins into the underground centuries before. The "unnatural" subterranean conditions have wrought even greater changes on the animals than on their masters (p. 101). Having moved from "the upper regions of light into the lower regions of darkness", their bodies have undergone "the most abnormal developments", namely, they have become more human in their appearance (pp. 100, 101). This by no means constitutes an improvement, but instead renders them "horrible", "ludicrous", "grotesque", "hideous", and "subnaturally" ugly (pp. 99-101).

Throughout the text, then, the detrimental effects of subterranean existence are made clear: for humans, to go down is to go back towards the animal, to regress, while for animals, descent generates an unnatural, overly-rapid evolution. The underground is, in other words, a place of topsy-turvy and inversion. That MacDonald is consciously engaging with contemporary theories of evolution is made clear, albeit in a rather ambivalent manner, in the text itself:

> [Curdie] had heard it said that they had no toes [...] he had not been able even to satisfy himself as to whether they had no fingers, although that also was commonly said to be the fact. One of the miners, indeed, who had had more schooling than the rest, was wont to argue that such must have been the primordial condition of humanity, and that education and handicraft had developed both toes and fingers – with which proposition Curdie had once heard his father sarcastically agree, alleging in support of it the probability that babies' gloves were a traditional remnant of the old state of things; while the stockings of all ages, no regard being paid in them to the toes, pointed in the same direction. (p. 58)[12]

---

reading, the underground enables the same kind of critique of contemporary society that is found in traditional katabatic narratives. In this context, see Clark, who refers to *katabasis* as an "instrument of national aspiration and social criticism" (Raymond J. Clark, *Catabasis: Vergil and the Wisdom-Tradition* (Amsterdam: Grüner, 1979), p. 14).

[12] MacDonald's awareness of, and participation in, the Darwinian debate is discussed in John Pennington, "Solar Mythology in George MacDonald's 'Little Daylight' and

Until the very last page of the text, the influence of the underground environment is exclusively negative, but it is at this point that MacDonald suddenly backs down, shying away from so unequivocal a position. Many goblins die in the flood they have themselves provoked, others escape and return to the surface, but some survive and remain underground. Of these, most "grew milder in character, and indeed became very much like the Scotch Brownies. Their skulls became softer as well as their hearts, and their feet grew harder, and by degrees they became friendly with the inhabitants of the mountain and even with the miners" (p. 458). Thus, the underground can foster physical and moral improvements as well as deterioration.

In a slightly later text, "The Day Boy and the Night Girl" (1879), MacDonald pursues this more positive perspective on the underground. Here, a witch embarks upon a programme of social engineering in which two children are brought up, from birth, in diametrically opposed conditions: a boy named Photogen is housed in a tower and never exposed to darkness, while a girl, Nycteris, dwells in an underground chamber modelled on an Egyptian tomb, and never experiences any light other than that of a dim lamp. Sixteen years spent underground have physical consequences for Nycteris which bring her dangerously close to the grotesque: "her optic nerves, and indeed her whole apparatus for seeing, grew both larger and more sensitive; her eyes, indeed, stopped short only of being too large".[13] When, with clear undertones of the Platonic cave narrative, she makes the difficult journey up to the surface, she is able to "see better than any cat", and to see colours invisible to ordinary human eyes (p. 182).[14] But her underground existence in no way leads to spiritual corruption or degeneration: she is sensitive and appreciative of the smallest things, impressively attuned to nature and – in her relationship with Photogen – shown to be sympathetic, compassionate and loving.

'The Day Boy and the Night Girl'", *Journal of the Fantastic in the Arts*, 10 (1999), 308-20.

[13] George MacDonald, "The Day Boy and the Night Girl", in *Victorian Fairy Tales: The Revolt of the Fairies and Elves*, ed. by Jack Zipes (New York/London: Routledge, 1989), pp. 175-208 (p. 179). Subsequent references are to this edition and are given after quotations in the main body of the text.

[14] See Frank Riga, "The Platonic Imagery of George MacDonald and C. S. Lewis: The Allegory of the Cave Transfigured", in *For the Childlike: George MacDonald's Fantasies For Children*, pp. 111-32.

Michael Mendelson has drawn attention to the fact that MacDonald has recourse to underground imagery in "very different contexts to very different thematic effects".[15] The two texts discussed above effectively illustrate the two basic positions available: either, as in "The Day Boy and the Night Girl", the underground has no detrimental effect or, as in *The Princess and the Goblin*, it leads to degeneration and regression – although not, as we have seen, of a necessarily definitive or absolute nature.

Like MacDonald, Jules Verne returned time and again to underground locales in his work.[16] Indeed, Rosalind Williams uses Verne to illustrate the difference between *katabasis* and the new form of narrative in which the underground is dwelling place rather than one-off destination: while *Voyage au centre de la terre* exemplifies the former, *Les Indes Noires* (1877) epitomizes the latter. At the beginning of this novel, just one family – the Fords – are known to live in the abandoned mine (the presence of their neighbours, the mad Silfax and his great-granddaughter Nell, is not revealed until much later).[17] But after the Fords discover a vast coal seam, they are joined by an entire community. Everyone who resides in the underground does so of his own free will, and, once in the underground, is unlikely to leave. A range of reasons for what may well seem a rather strange decision are offered in the course of the novel: not only do the inhabitants of Coal City escape the pollution and inclement climate of the Scottish surface, they also evade the tax man! Stress is also placed on the health benefits to be gained by living in "ce milieu parfaitement sain": Simon Ford states that the mine is superior to summer seaside resorts for those seeking cures, and maximizes the residents' prospects

---

[15] Michael Mendelson, "The Fairy Tales of George MacDonald and the Evolution of a Genre", in *For the Childlike: George MacDonald's Fantasies For Children*, pp. 31-49 (p. 40). Other texts by MacDonald involving undergrounds include *Phantastes* (1858), *The Golden Key* (1867), and *The Princess and Curdie* (1883).

[16] Other texts by Verne which prominently feature undergrounds, caves or caverns include *L'Ile mystérieuse* (1875), *Le Rayon Vert* (1882), *Sans dessus dessous* (1889), and *Face au drapeau* (1896).

[17] The relationship between Silfax and Nell is unclear. The narrator states that he is her "arrière-grand-père", but Nell later refers to herself as "la petite-fille du vieux Silfax" and to the latter as her "grand-père". Jules Verne, *Les Indes noires* (Paris: L'école des lettres/Seuil, 1993), pp. 295, 301, 303. Subsequent references are to this edition and are given after quotations in the main body of the text.

of longevity (pp. 72, 74).[18] The typical Vernian thumbnail sketches which present the characters emphasize their physical health and strength: Harry Ford, “un enfant de la houillère” whose entire existence “s’était écoulée dans les profondeurs de ce sol” (p. 50), is “un grand garçon de vingt-cinq ans, vigoureux, bien découplé” (p. 47); his mother is “grande et forte” (p. 73); his father “portait vigoureusement encore ses soixante-cinq ans” and is “[g]rand, robuste, bien taillé” (p. 68). Working and living underground like several generations of their ancestors before them has clearly had no adverse effect although, as in “The Day Boy and the Night Girl”, it has triggered physiological adaptation: not only does Harry have good hearing but he, like other characters of the text, also has excellent sight.

The discovery of Nell initially seems to call into question the wholly positive impact of the underground environment. Explicitly attributed to the “milieu exceptionnel” in which she had lived, Nell “paraissait n’appartenir qu’à demi à l’humanité. Sa physionomie était étrange” (p. 210). But Nell – “un être à la fois bizarre et charmant” – is, like Nycteris, supernatural fairy rather than subnatural monster (p. 209). Again like Nycteris, Nell initially cannot tolerate bright light, but her solitary subterranean life has had no permanent effect and she is soon fully integrated into Coal City’s healthy, industrious community. But what of Nell’s (great) grandfather? Knowledgeable and strong, despite his advanced age, Silfax is a crazed terrorist, a mad misanthrope who wages war against the new settlers. Williams refers to him as “a degenerate version of Nemo”, but is it underground conditions specifically which are responsible for this degeneration?[19] This certainly seems to be the case, and a degree of sympathy is aroused when Simon Ford states that it was his work as a “pénitent” clearing firedamp “[qui] avait dérangé ses idées” (p. 296). Yet it seems that some form of mental disorder predates his taking up of this role: Ford also states that Silfax had himself selected this dangerous role in accordance with his tastes, and soon afterwards asserts that his brain “a

[18] One naturally thinks of Corghan’s project in the Mammoth cave, a locale explicitly referred to in the course of *Les Indes noires*. Perhaps Verne was unaware that living underground harmed rather than helped the unwell: the health of Corghan’s patients soon deteriorated and some even died. The project was abandoned in 1843 and Corghan died from the disease six years later.

[19] Williams, *Notes on the Underground*, p. 171.

toujours été dérangé" (p. 297). What actually caused his madness, his degeneration, is never firmly established, but it seems clear that the underground environment is not responsible, especially given the robust physical and mental health of the other miners.

In both *Les Indes Noires* and *The Princess and the Goblin*, the underground is not only dwelling place but also workplace. In MacDonald's text, the goblins live in close proximity to the miners – including Curdie and his father, Peter – working alongside. The young hero's work is shown to be uncomfortable and cramped, undertaken in horribly warm conditions and exposed to the dangers of flooding in particular. Yet despite these factors, mining is explicitly described as being "not particularly unpleasant" (p. 48). Working in the "low and rather dreary earth" (p. 93) has had an impact upon Curdie's physical condition since the lack of light and air below ground has left his face "almost too pale", just as the eyes of Nycteris are almost too large (p. 36). Yet he is nevertheless "a very nice-looking boy" (p. 35) and, although somewhat limited in his imaginative and spiritual capacities at the start of the text, is certainly not in any way degenerate, like the goblin population against whom he is pitted.

Mining is also at the very heart of *Les Indes Noires*, yet Verne makes virtually no mention of the work undertaken in the underground. Coal City is much more a tourist destination, leisure park and housing estate than it is a workplace. As Marel observes: "On ne voit [...] jamais un groupe de mineurs 'haver' dans une taille. Dans cette immense fourmilière, tout semble se faire dans la facilité."[20] When the mining process is briefly described at the beginning of Chapter Nineteen, active human subjects are almost entirely eradicated through the use of the passive and of impersonal pronouns, as well as the focus on the mechanical:

> Ce jour-là, dans la Nouvelle-Aberfoyle, les travaux s'accomplissaient d'une façon régulière. On entendait au loin le fracas des cartouches de dynamite, faisant éclater le filon carbonifère. Ici, c'étaient les coups de pic et de pince qui provoquaient l'abattage du charbon; là, le grincement des perforatrices, dont les fleurets trouaient les failles de grès ou de schiste. Il se faisait de longs bruits caverneux. L'air aspiré par les machines fusait à travers les galeries d'aération.

[20] Henri Marel, "Jules Verne, Zola et la mine", *Les Cahiers Naturalistes*, 54 (1980), 187-200 (p. 196). This is one of the reasons why Marel concludes that "bien que l'action se déroule sous terre au milieu de la houille", *Les Indes Noires* does not constitute "un roman de la mine" (pp. 187, 198).

> Les portes de bois se refermaient brusquement sous ces violentes poussées. Dans les tunnels inférieurs, les trains de wagonnets, mus mécaniquement, passaient avec une vitesse de quinze milles à l'heure, et les timbres automatiques prévenaient les ouvriers de se blottir dans les refuges. Les cages montaient et descendaient sans relâche, halées par les énormes tambours des machines installées à la surface du sol. Les disques, poussés à plein feu, éclairaient vivement Coal-city. (pp. 279-80)

Both the engineer, James Starr, and Harry Ford refer to the work accomplished by the miners as being "dur" (pp. 13, 51), and the older man later speaks of the various risks faced in the mine: "le danger des éboulements, des incendies, des inondations, des coups de grisou qui frappent comme la foudre!" (p. 51) Yet for both Harry and Starr it is precisely these dangers which make the work interesting and appealing: "C'était la lutte, et, par conséquent, la vie émouvante!" (p. 51) Moreover, with an interesting play on words given the mining context, the narrator later makes clear that "l'homme, au fond, aime sa peine" (p. 71). In the course of the text, however, virtually no such dangers make themselves felt; the deaths or accidents which do occur are the direct result of Silfax's malevolent actions. Not only are there no real disadvantages associated with underground labour, but various factors in its favour are also outlined: for example mining, unlike agriculture, is both regular and lucrative. If, then, MacDonald is rather more ambivalent about *life* in the underground than is Verne, it is clear that *working* in the underground is not a negative experience in the text of either writer.

But in English and French texts for young readers which do focus on mining as an occupation, the underground is clearly a perilous place which causes deaths, accidents and physical debilitation. One of the most memorable, popular and critically acclaimed episodes in Hector Malot's *Sans Famille* (1878) is the hero's visit to the mining community of Varses in the Cevennes. Six weeks before Rémi's arrival in Varses an explosion had killed about ten men, and, on entering the town, his first encounter is with the deranged widow of one of the victims. Rémi's friend, the *magister*, whose hand was crushed in the mine when he was a young man, lives with the widow of a miner killed in an accident. It is, in similar fashion, because of an injury to the hand of Rémi's friend, Alexis, that the young hero himself temporarily takes up the mantle of miner. As he is working, the mine is flooded, and although after two weeks trapped below ground, Rémi himself escapes, almost 150 miners are killed. Deaths and accidents in the course of underground labour are also both portrayed and alluded to in A. de Gériolles's reworking of *Germinal* for young readers, *Sous Terre* (1910).[21] When a tunnel caves in midway through the text, one miner is killed and two are injured, while at the end of the novel, the flooding of the mine kills forty-eight men. As a result of his wife's earlier death by drowning in the mine, one of the secondary characters, Rabier, has been driven to drink so that mining accidents are shown to have moral repercussions. The underground is also a place where lives are lost and injuries sustained in R. M. Ballantyne's *Deep Down: A Tale of the Cornish Mines* (1868). In addition to the various references to accidents and deaths, the narrative also includes two explosions which blind or kill the workers involved. As in both *Sans Famille* and *Sous Terre*, there is also a flood, a "terrible catastrophe" that causes several deaths, five of which are particularized and reported directly.[22] Many other deaths and near-fatal incidents arising from the negligence or foolhardiness of the miners are also reported.

If all of the texts highlight the occupational hazards of mining, there is far less in the way of consensus as to the nature of underground working conditions and the long-term effects of the underground environment on the physical health of the miners. In *Sous Terre*, mining work is as horrific as in *Germinal*. Working in sodden clothes, in cramped conditions and extreme temperatures, with bleeding feet and knocks to the head, all the time breathing in coal-dust, this is truly an "existence d'enfer" (p. 45).[23] The effects of underground labour are wholly debilitative: the miners are thin, pale and "voûté" although not old, their faces "creusée[s] par la fatigue", (p. 33) their legs unnaturally swollen (p. 46). They develop coughs, angina and bronchitis. Similarly in *Deep Down*, underground labour, which takes place in high temperatures and bad air, is "toilsome" (p. 118). The local doctor states that such conditions are by no means

[21] For the text's intertextual status, see Kiera Vaclavik, "'Un Petit Costume de Mineur': Class and Gender Cross-Dressing in a Reworking of *Germinal* for Young Readers", *Romance Studies*, 21 (2003), 115-26.

[22] R. M. Ballantyne, *Deep Down: A Tale of the Cornish Mines* (London: Thomas Nelson, n.d.), p. 237. Subsequent references are to this edition and are given after quotations in the main body of the text.

[23] A. de Gériolles, *Sous Terre* (Paris: Hachette, 1923), p. 45. Subsequent references are to this edition and are given after quotations in the main body of the text.

conducive to health and various illnesses and symptoms are evoked including colds, inflammation of the lungs, and the spitting of blood. Few miners have the "health or strength" (p. 216) to go on working in the mine after the age of forty-five, and even young men have "hollow cheeks and bloodless lips" and an unshakeable cough (p. 381). Yet underground labour is also repeatedly shown to bring out the very best of the male body, as is clear from the following description of the miners, referred to as "Herculean men" (p. 342), going to collect their pay:

> There was a free-and-easy swing about the movements of most of these men that must have been the result of their occupation, which brings every muscle of the body into play, and does not – as is too much the case in some trades – over-tax the powers of a certain set of muscles to the detriment of others. (p. 381)

Descriptions of the miners abound with adjectives such as lean, wiry, robust, muscular, and athletic. The splendour of the male body in its prime is only emphasized by the environment:

> Here two men were "driving" the level, and another – a very tall, powerful man – was standing in a hole driven up slanting-ways into the roof, and cutting the rock above his head. His attitude and aspect were extremely picturesque, standing as he did on a raised platform with his legs firmly planted, his muscular arms raised above him to cut the rock overhead. (p. 349)

This is a superb example of "muscular Christianity", especially given that the character in question is a preacher as well as a miner.

Finally, it is worth noting that the boys who work in the mine in *Deep Down* are described as "diminutive but sturdy little urchins, miniature copies of their seniors, though somewhat dirtier; proud as peacocks" (p. 120), and that the main child protagonists grow up into a "strapping youth" and a "lovely girl" respectively (p. 452). Ballantyne, it seems, cannot quite make up his mind about mining which is at once dangerous and debilitative, but also elicits fine qualities. Malot, for his part, is much clearer. Rémi's first sight of the miners returning home does emphasize their suffering ("Ils s'avançaient lentement, avec une démarche pesante, comme s'ils souffraient dans les genoux"),[24] but the work is not only not difficult (a point made several times), but, as in *Les Indes Noires*, has various advantages:

> Le métier de mineur n'est point insalubre, et, à part quelques maladies causées par la privation de l'air et de la lumière, qui à la longue appauvrit le sang, le mineur est aussi bien portant que le paysan qui habite un pays sain; encore a-t-il sur celui-ci l'avantage d'être à l'abri des intempéries des saisons, de la pluie, du froid ou de l'excès de chaleur. (p. 52)

The illnesses, confined to a subclause, are easily missed and never expanded upon. Here, it is the surface and not the underground where the heat is problematic. The fact that Rémi decides not to pursue a career in mining is for temperamental and personal, not to mention narrative, reasons (the story must go on), not because mining is a debilitative, degeneration-inducing occupation to be avoided at all costs.

As in *Sans Famille*, where Rémi eventually ends up *en famille* in the upper echelons of English society, each of the mining narratives, as well as *Les Indes Noires*, *The Princess and the Goblin* and "The Day Boy and the Night Girl", have happy endings. This is even the case in *Sous Terre*, where, as we have seen, mining receives a wholly negative portrayal. Here, underground labour is simply left behind: the central family are plucked from the mine and set up in a grocer's shop by a benevolent fairy grandfather figure. For his part, the hero retains a connection with the mine since he becomes an engineer, but he spends little time underground. Indeed, his visit to the underground towards the end of the text, during which the mine is flooded and he is trapped, is undertaken simply in order to humour his friend (and future wife), Marthe.

Underground labour may be presented as being difficult and dangerous in these texts, but, with a couple of easily overlooked exceptions (such as Rabier's alcoholism), it has absolutely no moral repercussions. What a difference from Zola's portrayal of Jeanlin Maheu in *Germinal*, a novel which is intertextually related in various ways to several of the texts discussed in this article.[25] The pernicious

[24] Hector Malot, *Sans Famille*, 2 vols (Paris: Gallimard, 1997), II, p. 44. Subsequent references are to this edition and are given after quotations in the main body of the text.

[25] Zola, Verne and Malot all drew extensively on Simonin's *La Vie Souterraine* (1867) (see Henri Marel, *Germinal: Une Documentation intégrale* (Glasgow: Univer-

effect of underground labour on Jeanlin's body and soul can already be seen in the first part of Zola's novel: his simian nature is suggested when he is first introduced, he is shown engaging in sexual activities, stealing and drinking alcohol, and is later referred to as "un avorton humain, qui retournait à l'animalité d'origine" (p. 181).[26] But it is after his legs are crushed in a mining accident that the process of regression is accelerated: "[Etienne] le regardait, avec son museau, ses yeux verts, ses grandes oreilles, dans sa dégénérescence d'avorton à l'intelligence obscure et d'une ruse de sauvage, lentement repris par l'animalité ancienne. La mine, qui l'avait fait, venait de l'achever, en lui cassant les jambes" (p. 263). Jeanlin is, then, utterly different from the boys strutting around the mine and on their way to becoming "strapping" lads in Ballantyne's *Deep Down*. From his non-coincidentally subterranean refuge in part of the former workings on the mine, Jeanlin embarks on a wide-ranging campaign of theft. His rampages across the local vicinity eventually culminate in his cold-blooded and wholly unmotivated murder of the young soldier guarding the mine. In works for children, even where working conditions are poor or where the perils of mining are clear, there is none of the sense of outrage or protest discernible in *Germinal*. The endings serve to confirm the political conservatism, the respect for the status quo which characterizes much children's fiction.[27] The endings also convey the optimism typical of children's literature, and the texts overall powerfully promote the possibility of self-determination. In each of these texts, where time is spent underground either as inhabitants or workers or both, it is not the environment which shapes the individual. It is instead the individual's own behaviour and actions which determine his future. At most, the underground provides a space in which the heroes can demonstrate their bravery, quick-thinking and physical prowess. In other words, there is no danger that the heroes of these

sity of Glasgow French and German Publications, 1989). Like Zola, Verne also visited the mining community of Anzin. As mentioned above, *Sous Terre* can be seen as a rewriting of *Germinal* for young readers.

[26] Emile Zola, *Germinal* (Paris: Livre de Poche, 1983), p. 181. Subsequent references are to this edition and are given after quotations in the main body of the text.

[27] According to Caradec, it is because of the need to please parents as well as children that "[l]a littérature pour enfants est [...] la plus conventionnelle qui soit", and demonstrates an "attitude réactionnaire et conservatrice". See François Caradec, *Histoire de la littérature enfantine en France* (Paris: Albin Michel, 1977), p. 17.

texts – or even their villains, like MacDonald's goblins – will end up as Morlocks.

Alfred Jarry's Neo-Science: Liquidizing Paris and Debunking Verne

Ben Fisher

Abstract: *In his* Gestes et opinions du docteur Faustroll, pataphysicien: roman néo-scientifique, *Alfred Jarry combines scientific exposition and avant-garde artistic experiment. The novel inverts and perverts the physical and metaphysical universes in the course of an imagined voyage around Paris. This chapter maintains that, far from this being a simple artistic conceit, Jarry's work engages in a close and ingenious reading of contemporary science, including the earth sciences, offering in the process an avant-garde challenge to Verne's portrayal of geo-physical and geological phenomena.*

The themes proposed for examination in the conference "Histoires de la Terre" included "Tourism and the exploration of sublime landscapes", and "Creation or disappearance of islands". This essay discusses a text which exemplifies both headings, but in a most unusual manner; its landscapes are a set of islands created for the purpose of its tourism, and their creation is an artistic one, directly inspired, in turn, by the artistic creation of the figures who are represented as the lords and kings of their own extraordinary and utterly individual lands. While this Parisian voyage almost falls into the category of "travelling without moving", it also approaches and tests the limits of the boundaries between physical and conceptual universes. It has further relevance in that the novel in question includes an enigmatic critique of received ideas by the end of the nineteenth century about not only how the physical nature of the world may be perceived, but also how mankind should be "scientifically" educated to perceive and interact with it.

The text concerned is a novel, widely and not unreasonably regarded as "difficult",[1] by Alfred Jarry (1873-1907), an author whose notorious eccentricity still tends to obscure the remarkable breadth of an *œuvre* of which the scandalous play *Ubu Roi* is only a small and atypical element. The novel *Gestes et opinions du docteur Faustroll, pataphysicien, roman néo-scientifique* remained unpublished in full (the text may not be quite complete) until four years after Jarry's death, but most of the chapters relevant to this essay were published in the *Mercure de France* in May 1898, not long after the completion of the first manuscript (of two) of the novel. *Faustroll* represents Jarry's most sustained presentation and demonstration – both theoretical and practical – of his invented science of pataphysics. It is the practical demonstration that is of particular interest here.

The coherency and intended clarity of the demonstration of pataphysics in *Faustroll* do need to be stressed, as the science's meaning and application were by no means clear beforehand, and have remained highly flexible since. In particular the Collège de 'Pataphysique has developed concepts of pataphysics over several decades which have validity and currency in their own right, but would now be best described as parallel to Jarry's concepts. Within his own work and development, the word existed before it was given a firm meaning; its obscure origins lie in the Rennes schoolboy material (an extended tradition to which Jarry was only one of a number of contributors) which became the bulk of the *Ubu* plays; for example in Act II Scene 3 of *Ubu Cocu* (a play little altered from its schoolboy form) Père Ubu declares that: "La pataphysique est une science que nous avons inventée et dont le besoin se faisait généralement sentir", without the slightest further explanation (I, p. 497).[2] In Jarry's first novel, *Les Jours et les nuits* (1897), the chapter "Pataphysique" describes the central character Sengle's mental state, in which there is no distinction between thought and act, or between sleep and conscious-

[1] See for instance Noël Arnaud's preface to his edition of Alfred Jarry, *Gestes et opinions du Docteur Faustroll, pataphyscien*, (Paris: Gallimard, 1980), p. 7.

[2] All references to Jarry texts are to Alfred Jarry, *Œuvres complètes*, 3 vols (Paris: Gallimard/Pléiade, 1972-88). Subsequent references to this edition are given after quotations in the text.

ness, and in which demonstrable (if limited) powers of telekinesis are linked to Sengle's vision of the world as a boat with himself as the helmsman (I, pp. 793-5). This is consistent enough with the more explicit exposition of pataphysics in *Faustroll*, and it is also worth noting the following reservations for later reference:

> Et quant à sa vie pratique, il [Sengle] avait sûre confiance, ayant expérimenté toujours, *à moins que le principe de l'induction ne soit faux, mais alors les lois physiques seraient donc toutes fausses aussi*, qu'il n'avait qu'à s'en remettre au bienveillant retour des Extérieurs […] (I, p. 794; my italics)

In both of the works cited above, "la pataphysique" is a more or less incidental element in the text as a whole. In *Faustroll* it is brought into the foreground and "explained" early in the novel, but is also a pervasive presence throughout; this latter aspect is not so widely recognized, which is understandable given the more than eclectic character of the text. Thus the general definition of pataphysics is well known, and is the essence of the physicalized manifestations of the science in the section of the novel on which I shall concentrate: the pataphysical voyage around Paris, in which the imaginary and the symbolic become "real": "La pataphysique est la science des solutions imaginaires, qui accorde symboliquement aux linéaments les propriétés des objets décrits par leur virtualité" (I, p. 669).

While *Les Jours et les nuits* is played out within the realistic confines of Bohemian Paris, a provincial army barracks, and a military hospital in the capital, the voyage of Dr. Faustroll presents a Paris which, on first impression, appears entirely fantastical. Ostensibly, the voyage remains almost entirely within the physical confines of Paris. Given that pataphysics, as defined above, allows the imaginary to become real, these confines can include not only writers and artists, etc., but also intrusions of elements of their work into the world of the pataphysician's voyage, which starts in the conventionally real world, where Faustroll is evicted from his home in the rue Richer. The pataphysical concepts of dimensions and existence are limited and defined only by imagination, and are thus remarkably flexible. The voyage is best known for its literary and artistic aspects, but also has conceptual and methodological debts to physical science; both sides are used to distort and redefine perceptions.

Dr. Faustroll and his two companions, the bailiff narrator Panmuphle and Bosse-de-Nage, a monkey of few words (limited to "ha ha", a few Belgian words aside [I, p. 672]), travel by boat – a twelve-metre elongated sieve coated with paraffin, which "floats" through surface tension effects:

> Mon crible flotte donc, à la manière d'un bateau, et peut être chargé sans couler à fond. Bien plus, il possède sur les bateaux ordinaires cette supériorité, m'a fait remarquer mon savant ami C.-V. Boys, qu'on peut y laisser tomber un filet d'eau sans le submerger. Que j'expulse mes urates ou qu'une lame embarque, le liquide passe à travers les mailles et rejoint les lames extérieures. (I, p. 664)

Charles Vernon Boys is best known for, precisely, his work on effects related to surface tension. These were widely disseminated in his *Soap Bubbles: Their Colours and the Forces Which Mould Them* of 1890, which Jarry follows closely. Jarry also makes reference to Boys's technique of using a crossbow to produce quartz fibre, from which Faustroll has made another boat. It is however in the floating sieve (known as the *as*) that he sets off on his voyage, having been forced to leave his home upon his eviction by Panmuphle:

> Je suis d'autant mieux persuadé de l'excellence de mes calculs et de son insubmersibilité, que, selon mon habitude invariable, nous ne naviguerons point sur l'eau, mais sur la terre ferme. (I, p. 665)

Panmuphle narrates their departure thus: "Nous nous insérions entre les foules d'hommes ainsi que dans un brouillard dense, et le signe acoustique de notre progression était celui de la soie déchirée" (I, p. 675). The section of the journey which interests us most here is the first, Livre III, with its obvious debt to a distinctively Rabelaisian form of the literary voyage (indeed the name of the penultimate island is a borrowing from the *Cinquième Livre*). The chapters are dedicated to two publishers, three artists, one composer and eight writers, all of them friends or acquaintances of Jarry's (dedications noted in square brackets in the following list).

Livre III
De Paris à Paris par mer, ou le Robinson belge

XI. De l'embarquement dans l'arche [Alfred Vallette]
XII. De la mer d'Habundes, du phare olfactif, et de l'île de Bran, où nous ne bûmes point [Louis Lormel]
XIII. Du pays de Dentelles [Aubrey Beardsley]

XIV. Du bois d'Amour [Émile Bernard]
XV. Du grand escalier de marbre noir [Léon Bloy]
XVI. De l'île amorphe [Franc-Nohain]
XVII. De l'île fragrante [Paul Gauguin]
XVIII. Du château-errant, qui est une jonque [Gustave Kahn]
XIX. De l'île de Ptyx [Stéphane Mallarmé]
XX. De l'île de Her, du cyclope et du grand cygne qui est en cristal [Henri de Régnier]
XXI. De l'île Cyril [Marcel Schwob]
XXII. De la grande église de Muflefiguière [Laurent Tailhade]
XXIII. De l'île sonnante [Claude Terrasse]
XXIV. Des ténèbres hermétiques, et du roi qui attendait la mort [Rachilde].[3]

The boundaries of Paris are stretched somewhat to accommodate certain islands. The "bois d'Amour" represents the artists' colony at Pont-Aven, closely associated with the Nabi group with whom Jarry was friendly,[4] and "du château-errant" represents Gustave Kahn's holiday retreat in Belgium – or as the text has it, the north-east of Paris (I, p. 684).

Travelling by boat on solid land, it is only logical that the islands should be liquid:

> La surface de l'île (il était naturel que les îles nous parussent comme des lacs, en notre navigation de terre ferme), est d'eau immobile, comme d'un miroir [...] (I, p. 686, from the chapter dedicated to Henri de Régnier)

This is not however an absolute constant, as the islands are adapted to the images that best suit their subjects; what unites the islands is that they are all brought into a dimension where the reader can conceive them physically and visually; the voyage has even been transformed into an illustrated board game.[5] The islands are complex, microcosmic artistic constructs, which in most cases combine imagery from an au-

---

[3] Basic introductions to the figures visited and the imagery used can be found in the apparatus of the Pléiade *Œuvres complètes*. More extensive ones can be found in the Cymbalum Pataphysicum's rare 1985 edition of the novel; some care should be exercised as these are mostly drawn, without revision, from dated material published by the Collège de 'Pataphysique in the 1950s.

[4] On Jarry's links with Pont-Aven see Jill Fell, *Alfred Jarry: an Imagination in Revolt* (Madison & Teaneck: Fairleigh Dickinson University Press, 2005).

[5] "Le Strobile jeu de HA HA", by the Regent Gil of the Collège de 'Pataphysique, reproduced in: Collège de 'Pataphysique, *Les très riches heures du Collège de 'Pataphysique* (Paris: Fayard, 2000), p. 119.

thor or artist's work with more personal allusions to the hospitality of the ruler of each isle; the visits to almost all are celebratory and gregarious, with the exception of the scatological vilification of the publisher Louis Lormel embodied in the first island.[6] There appears to be no particular order within the visits to the islands, except that they are clearly framed within the literary and artistic orbit of the *Mercure de France*. The embarkation is dedicated to the review's co-founder and publisher Alfred Vallette, and the last island visited is ruled over by a king who represents his novelist wife Rachilde. At the time, the *Mercure* was still based in its original premises in the rue de l'Échaudé, approached via the boulevard St-Germain, which is transformed thus:

> Ayant passé le fleuve Océan, qui est fort analogue, pour la stabilité de sa surface, à une vaste rue ou boulevard, nous arrivâmes au pays des Cimmériens et des Ténèbres hermétiques, qui en diffère comme peuvent différer deux plans non liquides, par la grandeur et la division. (I, p. 693)[7]

There are functional variations in the systems of imagery from which the islands are constructed, determined by the images available, and with each variation having its own peculiar richness. The island dedicated to poet Gustave Kahn is a good example of one based on both artistic and personal references; the "château-errant, qui est une jonque" itself draws on Kahn's *Les Palais nomades* and maritime poems from other of his collections, but the landscape in which this strange ship appears, disappears and reappears is clearly that of Kahn's holiday retreat at Sint Anna on the Belgian coast, which Jarry had visited in 1896:

> Hespaillier infatigable, je tirai les avirons plusieurs heures, sans que Faustroll parût découvrir l'abord enfin proche du château fuyant selon des mirages; après des rues étroites de maisons désertes espionnant notre venue par les yeux à facettes de compliqués miroirs, nous touchâmes de la fragilité sonore de notre proue l'escalier de bois ajouré du nomade édifice. [...]
>
> Le palais était une bizarre jonque sur une eau calme ouatée de sable [...]

---

[6] The reasons for Jarry's rupture with Lormel (*pseud.* of Louis Libaude), who had published some of his earliest work, remain obscure.

[7] The "Ténèbres hermétiques" of the chapter are of course an allusion to Hermes/Mercury/*Mercure*, as well as to hermeticism.

> Dès que l'amarre eut été détachée par notre fadrin laconique, le château croula et mourut, et reparut miré dans le ciel, des lieues plus loin, la grande jonque éraillant le feu du sable. (I, pp. 684-5)

A different approach is visible in "Du pays de Dentelles", in which the king visited by Faustroll and his companions is Aubrey Beardsley; in this case the imagery is derived entirely from Beardsley's art, and there are no readily distinguishable real-life references.[8] For instance, having introduced the remarkably pure light of this island:

> Le roi des Dentelles l'étirait comme un cordier persuade sa ligne rétrograde, et les fils tremblaient un peu dans l'obscurité de l'air, comme ceux de la Vierge. Ils ourdirent des forêts, comme celles dont, sur les vitres, le givre compte les feuilles; puis une madone et son Bambin dans de la neige de Noël; et puis des joyaux, des paons et des robes, qui s'entremêlaient comme la danse nagée des filles du Rhin. (I, pp. 677-78)

While the details of Jarry's own dealings with Beardsley remain sketchy, it is established that they knew each other. Thus the lack of points of reference remains curious – all the more so as there is no explicit commemoration of Beardsley's death in March 1898, which falls within the possible dates of composition of the first manuscript of *Faustroll*, and certainly predates the second. We should also note Jill Fell's speculative but incisive reading of this chapter, which proposes a system of wordplay and concealed subtleties that might include reference to the recent Oscar Wilde scandal.[9]

Jarry also demonstrates the adaptability of a virtuoso when creating an island to honour a figure whose own work is not replete with the complex, suggestive imagery that typifies the late Symbolist environment. The writing of the fiery, uncompromising religious novelist and pamphleteer Léon Bloy is very much rooted in an all too real world characterized by grinding poverty. So in "Du grand escalier de marbre noir" (a punning allusion to Marchenoir, semi-autobiographical protagonist of Bloy's novel *Le Désespéré*), Jarry

---

[8] See J. Smaragdis (*pseud.*), "Du Pays des Dentelles" [*sic.*], *Cahiers du Collège de 'Pataphysique*, no. 22-23 (23 palotin 83 E.P. [dated by the Collège's "calendrier pataphysique"], i.e May 1957), 65-67.

[9] Fell, *Alfred Jarry: an Imagination in Revolt*, pp. 131-36.

creates imagery derived from Bloy's mission to convert his readership to his own form of ascetic, anti-clerical Catholicism:

> Au sortir de la vallée, nous longeâmes un dernier calvaire, que l'effroi de sa hauteur aurait permis de prendre, sans examen, pour un monumental autel de messe, noir. A la pointe mousse de l'impraticable pyramide de marbre obscur, entre deux acolytes bien semblables à des cynocéphales de Tanit, la tête du roi géant se carbonisait devant la fournaise de la lune. Il empoignait un tigre par l'extensibilité de la peau de son cou, et forçait le peuple de la mer d'Habundes à une ascension à genoux. (I, pp. 680-81)

The amorphous, ambiguous physicality of the voyage is nowhere clearer than in its most famous chapter, "De l'île de Ptyx", in which Faustroll and his companions visit Stéphane Mallarmé. Jarry also transformed the chapter into a "nécrologie" for Mallarmé in his 1899 *Almanach du Père Ubu*, incorporating Mallarmé's own enthusiastic reaction to the chapter upon its appearance in the *Mercure de France* (I, pp. 564-65).[10] "L'île de Ptyx est d'un seul bloc de la pierre de ce nom, laquelle est inestimable, car on ne l'a vue que dans cette île, qu'elle compose entièrement" (I, p. 685). The name of the island is drawn from the second quatrain of Mallarmé's famous "sonnet en – yx", quoted here in its original form of 1868 (the one dating from 1887 is significantly different but the point remains the same):

> Sur des consoles, en le noir Salon: nul ptyx,
> Insolite vaisseau d'inanité sonore,
> Car le maître est allé puiser de l'eau du Styx
> Avec tous ses objets dont le Rêve s'honore.

Jarry is also making use of the no less famous difficulty in finding any set meaning in what Mallarmé called its "mirage interne des mots mêmes" at the time of its composition;[11] the poem both invites and denies the possibility of interpreting the neologism "ptyx". Jarry none-

---

[10] Ben Fisher, *The Pataphysician's Library: an exploration of Alfred Jarry's "livres pairs"* (Liverpool: Liverpool University Press, 2001), p. 82. The letter adapted by Jarry described the published chapters as "du Rabelais, dira-t-on, mais ce que ce divin eût écrit originellement tout à l'heure" – Stéphane Mallarmé, *Correspondance*, 12 vols (Paris: Gallimard, 1959-98), vol. X (1984), pp. 190-91.

[11] See the letter of 18 July 1868 to Henri Cazalis which includes the sonnet. Stéphane Mallarmé, *Œuvres complètes*, 2 vols (Paris: Gallimard/Pléiade, 1998-2003), I, pp. 730-33.

theless describes this elusive, impossible but essential item as an object which is simultaneously material and abstract:

> On n'y percevait plus les accidents des choses, mais la substance de l'univers, et c'est pourquoi nous ne nous inquiétâmes point si la surface irréprochable était d'un liquide équilibré selon des lois éternelles, ou d'un diamant impénétrable, sauf à la lumière qui tombe droit. (I, p. 685)

This also neatly illustrates a principle which is a feature of Jarry's definition of pataphysics earlier in the novel, namely a proto-Cubist notion that the characterization or indeed definition of a physical thing is not fixed. It depends entirely on perspective, or rather the acceptance of multiple, simultaneous perspectives – a principle which is made concrete (in its solid or liquid form) in the pataphysical voyage.

> Pourquoi chacun affirme-t-il que la forme d'une montre est ronde, ce qui est manifestement faux, puisqu'on lui voit de profil une figure rectangulaire étroite, elliptique de trois quarts, et pourquoi diable n'a-t-on noté sa forme qu'au moment où l'on regarde l'heure? Peut-être sous le prétexte de l'utile. (I, p. 669)

And however intensely hyper-artistic the expression of this may be, the way in which it is narrated is more communicative than it may appear at first sight, and the above reflection about the shape of a watch also conveys a definite sense of investigation, as well as of frustration with two-dimensional thinking. Bringing these aspects together, if we were to attempt a more prosaic analysis than this novel usually attracts, we might observe that there is a strong pedagogical streak, albeit a highly unconventional one given the complex universe being unveiled within and alongside the apparently real one. As a narrator Panmuphle is anything but omniscient; he is very much the neophyte, representing the reader, filled with a wonderment at the supplementary universe which Dr Faustroll reveals and explains.

This question of pedagogy is where it becomes appropriate to introduce Jules Verne – or more precisely, Jarry's subtle but damning critique of the *roman scientifique* of which Verne remains the pre-eminent practitioner. Jarry actively invites the reader to consider Verne by including *Le Voyage au centre de la Terre* (1864), as the last of the pataphysician's twenty-seven "livres pairs" – volumes seized by Panmuphle, and reduced to essences (the "petit nombre des élus") by

Faustroll so that they may be taken along on the voyage.[12] Among the thirteen living authors whose books are selected (five of whom are visited on Faustroll's voyage), Verne is the only one with no links to Jarry's own artistic circles.[13] It may appear easy to assume that *Le Voyage au centre de la Terre* is included as a childhood favourite; the very wide range of books within the list (from *The Odyssey* to the latest avant-garde poetry and prose) does include a small number of other texts which Jarry may or may not have known in childhood, such as a harlequinade by Jean-Pierre Claris de Florian and a children's tale by Marceline Desbordes-Valmore. There is, however, a total lack of any corroborative evidence for this. Throughout his work, Jarry is the most allusive of authors, and his reading is relatively easy to trace, regardless of its diversity and frequent obscurity. Yet his only other reference to Verne is an offhand one ("Jules Verne en écrirait un roman") in a humorous article written for *Le Canard sauvage* in 1903 (II, p. 468). In this light, Verne's presence among Faustroll's books would not appear to be a random item in isolation – in fact it sticks out like a sore thumb.

I believe it is a provocative choice, not least because *Gestes et opinions du docteur Faustroll* includes a substantial apparent borrowing from *Le Voyage au centre de la Terre*. In both voyages, there are three travellers, and Jarry appears to base his unusual trio on Verne's.[14] Otto Lidenbrock and Faustroll are both explorers of inner space; Faustroll's exploration of it is not limited to the artistic constructs of the sea of Paris, but also includes (prior to the start of the main voyage) a brief excursion to explore the element of water, by miniaturizing himself in order to better examine a raindrop ("Faustroll plus petit que Faustroll", I, pp. 670-71). The narrators Axel Lidenbrock and Panmuphle are both neophytes learning from both the words and the examples of the scientists, and they also share a lack of depth in their characterization. And in both novels the physical hard work is done by a character of very few words; as already remarked,

[12] I have explored these volumes and their relevance to Jarry extensively in *The Pataphysician's Library*.

[13] The thirteen are Léon Bloy, Georges Darien, Max Elskamp, Gustave Kahn, Maurice Maeterlinck, Stéphane Mallarmé, Catulle Mendès, Joséphin Péladan, Rachilde (under her additional pseudonym "Jean de Chilra"), Henri de Régnier, Marcel Schwob, Émile Verhaeren, and Verne. The death of Mallarmé in September 1898 would reduce the number to twelve between the two manuscripts of *Faustroll*.

[14] Fisher, *The Pataphysician's Library*, p. 124.

the monkey Bosse-de-Nage's practical vocabulary is limited and repetitive:

> "Ha ha!" disait-il en français; et il n'ajoutait rien davantage. (I, p. 672)
>
> "Ha ha!" dit-il compendieusement; et il ne se perdit point dans des considérations plus amples. (I, p. 678)
>
> "Ha ha!" dit-il, mais nous n'écoutâmes point la suite de son discours. (I, p. 684)

His earlier parallel in Verne is the Icelander Hans, "personnage grave, flegmatique et silencieux";[15] it is perhaps as well that he says very little, as it has been shown that what Verne presents as Icelandic or Danish vocabulary is in fact Swedish.[16]

Whatever its enduring appeal as an adventure, in scientific terms *Le Voyage au centre de la Terre* gives every appearance of being intended mainly for the young, to instil scientific knowledge and in particular a sense for investigative deduction; its status as a children's book has attracted serious academic discussion.[17] Thus however complicated the Lidenbrocks' journey may appear to be, from their descent in Iceland to their re-emergence among the lava of Stromboli, at heart the novel has a straightforward linear plot where every mystery has a solution; even the deciphering of the clues in Arne Saknussemm's runic manuscript is an essentially methodical process in which one step follows logically upon another.[18] These directions allow Verne's three explorers to enter the deepest bowels of the Earth, even if they never actually reach the centre of the globe. But in terms of scientific method, reason conquers all.

Jarry does not believe in inductive reasoning of this kind. The doubt expressed in *Les Jours et les nuits* has become all but an explicit refutation in *Faustroll*:

> La science actuelle se fonde sur le principe de l'induction: la plupart des hommes ont vu le plus souvent tel phénomène précéder ou suivre tel autre, et en concluent

[15] Jules Verne, *Le Voyage au centre de la Terre* (Paris: Livre de Poche, n.d.), p. 93.
[16] Daniel Compère, *Un Voyage imaginaire de Jules Verne: "Voyage au centre de la Terre"* (Paris: Minard, 1977), pp. 50-51.
[17] Isabelle Jan, "*Le Voyage au centre de la Terre* est-il un livre pour enfants?", in Colloque d'Amiens [11-13 novembre 1977], *Jules Verne, écrivain du XIXe siècle*, 2 vols (Paris: Minard, 1978), vol. 1, pp. 81-88.
[18] Verne, *Le Voyage au centre de la Terre*, pp. 9-40.

qu'il en sera toujours ainsi. D'abord ceci n'est exact que le plus souvent, dépend d'un point de vue, et est codifié selon la commodité, et encore! (I, p. 669)

Furthermore, by populating Dr Faustroll's boat with characters who are, among other things, structured parodies of Verne's explorers, he enhances the implied critique within the text, as the marvels revealed in Faustroll's voyage remain within the realm of the marvellous, and are explained (insofar as they are explained) through complex artistic reference rather than scientific method.

However we are not dealing with any kind of general rejection of science within the *Gestes et opinions*. In addition to its artistic references, Jarry's novel also demonstrates an unusually keen interest in a range of contemporary or recent scientists – many of them British – with a particular emphasis on ones famous for success through practical experiment as much as through theory. For details of these references and how Jarry may have become familiar with them, Linda Klieger Stillman's investigation is thoroughly recommended.[19] Charles Vernon Boys has already been mentioned, and other prominent examples include Michael Faraday and William Crookes; the chapter "Faustroll plus petit que Faustroll" is dedicated to the latter. While Jarry's immediate reference to the physicist and chemist Crookes is to the method of a paper Jarry is likely to have read in the *Revue scientifique* in 1897 (see notes, I, p. 1223), it is interesting to speculate (there is no evidence) whether Jarry was familiar with the more eccentric side of Crookes's work, as it could be relevant to later parts of the novel. Crookes investigated telepathy, which Jarry uses as a device; his experiments into spiritualism may also have some relevance as they led him to place faith in it after initial scepticism.[20] I suggest this may be relevant because while Panmuphle is the only one of the voyagers to physically survive, both Bosse-de-Nage and Faustroll's participation in the novel is undiminished by their deaths.

And science is very much to the fore after Faustroll leaves the Earth by way of the Paris Morgue. From the realm of "Éthernité", he communicates by telepathic letters to Lord Kelvin (I, pp. 724-28), first explaining his attempt to establish fundamental physical measurements in terms drawn from the published lectures and speeches of the

[19] Linda Klieger Stillman, "Physics and Pataphysics: the Sources of *Faustroll*", *Kentucky Romance Quarterly*, 26 (1979), 81-92.
[20] http://en.wikipedia.org/wiki/William_Crookes (accessed 25 March 2007).

great Scottish scientist. Faustroll then uses these fundamentals to establish that the Sun is cold,[21] and with the additional support of algebra, proceeds to measure the surface of God. Jarry would refer again to Kelvin's lectures in an article for *La Plume* in 1903, in which he claimed, perhaps eccentrically, that the *roman scientifique* was a direct descendant of alchemical tales from the *Mille et Une Nuits*, and he held up H.G. Wells as the modern master of the genre without mentioning Verne at all – an omission which seems bold enough to be deliberate (II, pp. 519-20).

Jarry is a hugely eclectic writer (indeed this is a significant part of his charm), and *Gestes et opinions du docteur Faustroll, pataphysicien* is a highly experimental text. It does however bear examination as a sustained attempt to justify its subtitle of "roman néo-scientifique", through if not a fusion then certainly a juxtaposition of what were then "advanced" figures in both science and the artistic avant-garde. Jarry's fantastical perspective of a Paris distorted according to artistic principles serves as an appropriate parallel to the work of scientists who – most prominently Kelvin – were coming to question the inherited limits and certainties of their sciences, just as much as the writers and artists of *fin-de-siècle* Paris were challenging the conventions of their own disciplines.

[21] This is a potential but unstated parallel with Verne, who maintains through the "science" of *Le Voyage au centre de la Terre* that the core of the Earth is cold.

SECTION 4

EARLY TWENTIETH CENTURY

Reading Environmental Apocalypse in J.-H. Rosny Aîné's Terrestrial Texts

Louise Lyle

Abstract: *Technological advances in exploiting the earth's resources were not uncritically welcomed by all. This chapter offers an analysis of J.-H. Rosny Aîné's development of the notion of evolutionary narratives inscribed upon the terrestrial text of the planet itself, through which the author may be seen to anticipate an "apocalyptic environmentalist" perspective on mankind's increasingly exhaustive exploitation of the Earth.*

Belgian brothers Joseph-Henri (1856-1940) and Séraphin-Justin (1859-1948) Boëx began their long and prolific literary careers under the collective pseudonym of J.-H. Rosny in the 1880s. Writing first together, then individually under the respective pseudonyms of J.-H. Rosny Aîné and J.-H. Rosny Jeune, they produced an extensive *œuvre* comprising poetry, theatre, press articles, essays and works of scientific popularization. It is, however, for their novels and short stories, falling variously into the realist, philosophical and fantastical genres,[1] that they are best remembered nowadays. In this chapter, I will consider J.-H. Rosny Aîné's[2] development of the concept of the earth in certain of these latter works as a kind of text in which mankind's apocalyptic destiny may be read.

---

[1] Gérard A. Jaeger, *Approche critique et bibliographique des frères Rosny* (Sherbrooke: Éditions Naaman, 1986), p. 11.

[2] For brevity, the elder of the brothers, whose works are the focus of this chapter, will hereafter be denoted simply as Rosny.

In *La Charpente* of 1900, Rosny's protagonist, a socially progressive geographer and sociologist named Duhamel, rejects the notion of "un monde immobile et définitif" in favour of "un monde relatif et en mouvement", describing "la croûte terrestre" as "[u]n livre [qui] nous a été laissé, plus vaste que la Bible", in which "il s'y trouve écrit une vérité qui a des millions d'années: l'évolution, le perfectionnement des organismes".[3] Serving not only, therefore, as a vast source book of palaeontological inspirations for the meticulously described primordial landscapes and exotic flora and fauna for which Rosny's prehistoric novels were much admired,[4] planet earth is also, and even more significantly from our point of view, seen by the author as a developing record of processes that are still on-going. For, if the earth is regarded as a text in which may be read the geological and biological changes to which it and its inhabitants, human or otherwise, have been and continue to be subject, evolution may be understood as the narrative linking a whole that is perpetually incomplete in respect of the chapters still to be written.

As Daniel Compère, comparing Rosny and H.G. Wells, explains: "Tous deux ont conscience que l'homme est une étape dans l'évolution, et c'est sur cette certitude que s'appuie leur vision de l'avenir. Tous deux suivent le même mouvement qui consiste à utiliser l'histoire, non plus pour expliquer le passé, mais pour décrire le futur."[5] It is, then, by means of the extrapolation of evolutionary dynamics inferred from the distant past that Rosny constructs his vision of futurity in the work which will be the central focus of this chapter, namely *La Mort de la Terre*, a short *roman d'anticipation* of 1908. Our examination of these evolutionary dynamics, and of the author's construction of the earth as the text in which human history is viewed as a progression spanning past, present and future will firstly, however, take account of two other works, *Les Xipéhuz* of 1887 and *Les Navigateurs de l'infini* of 1910, which may be classified with *La Mort*

---

[3] J.-H. Rosny, *La Charpente* (Paris: Éditions de la Revue Blanche, 1900), p. 259.
[4] These works include *Vamireh* (1891), *Eyrimah* (1895), *Elem d'Asie* (1896), *Nomai, Amours lacustres* (1897), *La Guerre du feu* (1911), *Le Félin géant* (1919), *Le Trésor dans la neige* (1921), *Les Conquérants du feu* (1929) and *Helgvor du fleuve bleu* (1930).
[5] Daniel Compère, "La Fin des hommes", *Europe*, no. 681-682, jan-fév 1986, 29-36, (p. 30).

de la Terre under the rubric of Rosny's *romans du merveilleux scientifique*.

Rosny's livre lapidaire

Les Xipéhuz of 1887 recounts the experiences of an early human civilization, situated in time and space "mille ans avant le massement civilisateur d'où surgirent plus tard Ninive, Babylone, Ecbatane".[6] The peace-loving, nomadic Zahelal people find themselves under threat from strange creatures, initially designated in the text as "les Formes", made of unidentified mineral substances and assuming regular geometrical forms. Unassuaged by the sacrifices offered up to them, these mysterious forms exercise complete dominance within the constantly expanding boundaries of their territory, effortlessly annihilating the warriors who seek to contain them, and instilling in the tribes, for the first time, a sense of their own fallibility: "*L'homme allait périr*. L'*autre*, toujours élargi, dans la forêt, sur les plaines, indestructible, jour par jour dévorerait la race déchue" (author's emphasis, p. 634).

Rosny thus posits a kind of eschatological consciousness as an almost innate feature of the human psyche, present virtually from the dawn of civilization, as if in anticipation of the millennialism of the Christian theologians and their spiritual descendants. Juxtaposing "l'an mil des peuples enfants, [...] [et] la résignation de l'homme rouge des savanes indiennes" (p. 634), Rosny views this innate anxiety about our final end as an omnipresent feature of human history, whether in respect of his own fictitious pre-Mesopotamian tribes as they fall prey to the enigmatic conical "Formes", or of the massacres of Native Americans at the hands of the armies descended from European colonizers which continued until the end of the nineteenth century. Rosny's evocation of humanity's consciousness of impending doom in both historical and contemporary contexts in *Les Xipéhuz* also anticipates his subsequent writings dealing with this same theme in futuristic settings, such as *Les Navigateurs de l'infini* and *La Mort de la Terre*.

6 J.-H. Rosny Aîné, *Les Xipéhuz. Romans Préhistoriques* (Paris: Robert Laffont, 1985), 627-52, p. 629. Subsequent page references to this text will be given in brackets after the relevant quotation.

Thematic continuity between the works is not, however, confined to man's growing awareness that his reign on earth will not be limitless, but also extends to the inscription of human history in the stone of which the earth itself is formed. In *Les Xipéhuz*, the desperate tribes turn for help to Bakhoûn, established by the author as a more highly intellectually evolved human being who has abandoned paganistic beliefs and nomadism in favour of a rationalistic philosophy and settled lifestyle. Entreated to save the tribes from the advancing menace which he names "les Xipéhuz", Bakhoûn undertakes not only to study the creatures in order to determine how best to contain them, but also to recount the episode in what will in subsequent centuries come to be recognized as "[un] grand livre antécunéiforme de soixante tables, le plus beau livre lapidaire que les âges nomades aient légué aux races modernes" (p. 635). It is from these stone tablets that the remainder of the account of mankind's first recorded battle for survival is drawn, via "la merveilleuse traduction de [l'illustre savant] M. Dassault", apparently converted, "[d]ans l'intérêt du lecteur [...] en langage scientifique moderne" (p. 636). Rosny thus contrives to remove what André Maraud calls "le double obstacle de la langue et du style", abolishing the comprehension gap between ancient text and modern reader while justifying "l'intervention de l'écrivain [...] par la nécessité de mettre à la portée du lecteur profane un texte dont la traduction littérale est d'un accès difficile".[7]

Bakhoûn's extensive observations of the Xipéhuz reveal that the area of the territory they occupy is directly proportional to their population size, leading him to reflect sadly on the impossibility of their coexistence with humans. Self-sufficient, ruthless and highly intelligent, the Xipéhuz are formed of an unknown mineral substance and have a small star-like formation at the base of their structure from which they project both the killer rays that incinerate their victims and the light by which they communicate with one another, as Bakhoûn explains:

> Supposons, par exemple, qu'un Xipéhuz veuille parler à un autre. Pour cela, il lui suffit de diriger les rayons de son étoile vers le compagnon, ce qui est toujours perçu instantanément. [...] Le parleur, alors, trace rapidement, sur la surface même de son interlocuteur [...] une série de courts caractères lumineux, par un jeu de

[7] André Maraud, "Le Texte à l'origine de l'histoire", *Europe,* no. 681-682, jan-fév 1986, 101-6, p. 105.

> rayonnement toujours émanant de la base, et ces caractères restent un instant fixés, puis s'effacent. (p. 639)

While the motif of writing on stone is evoked once again, the ephemeral character of the Xipéhuz script seems to hint at the corresponding impermanence of its scribes, for, as Maraud notes, Rosny's aim in this work is to exalt "la fonction proprement humaine de l'écriture, qui conserve la mémoire de l'espèce".[8]

Thus anticipated, the destruction of the Xipéhuz is inevitable once Bakhoûn has discovered their Achilles heel. Susceptible to attack at the luminous star formation which is the sole vulnerable point of their otherwise indestructible anatomy, the redoubtable creatures are reduced to mere rubble by the massed hordes of tribal warriors armed with arrows and spears during two days of intensive combat. The only reminders of their existence are Bakhoûn's narrative and the mysterious "cristaux jaunâtres, disposés irrégulièrement, et striés de filets bleus" (p. 637), conserved for posterity in a London museum, between which Maraud draws the following comparison:

> Histoire des origines, le livre lapidaire est aussi l'origine de l'histoire. […] Le texte oppose légitimement aux "débris minéraux", restes fossiles des "vivants" que furent les Xipéhuz, et qui se refusent à l'*analyse*, "les tables de granit" où Bakhoûn a inscrit leur histoire, véritables pierres vives qui constituent le premier des livres... [9]

Bakhoûn's writings are thus presented by Rosny as a fictitious exemplar of man's first *conscious* efforts to record his own history for presumed future generations. The sixty painstakingly inscribed stone tablets constituting this magnificent artefact dating from earliest antiquity are, however, dwarfed in scale and significance by the much greater *livre lapidaire* of the Earth. Truly "le premier des livres", the Earth may be understood as the vast and still unfinished multi-volume work from which these few stone pages have themselves been cut, in which a record of *all* living things throughout the ages is contained. It is to Rosny's rendering of the evolutionary narrative running through the entirety of this vast geological chronicle, within which the story of mankind is no more than a brief chapter, that we shall now turn.

---

[8] Maraud, "Le Texte à l'origine de l'histoire", p. 104.
[9] Maraud, "Le Texte à l'origine de l'histoire", p. 106.

The Struggle for Life

The extent to which Rosny drew inspiration from the evolutionary theories of Charles Darwin is apparent in numerous of his works. Social Darwinist attitudes are consistently condemned, whether, as Linda Clark suggests, in respect of race relations between the beleaguered Native Americans and European settlers in *Le Serment* of 1896,[10] or in relation to the unscrupulous suppression of rivals in politics and commerce in *La Charpente* of 1900.[11] Mélanie Bulliard, meanwhile, discerns a more ambitious purpose in Rosny's evolutionary enthusiasms. In *L'Enjeu des origines*, she proposes that the author's extensive scientific knowledge and naturalist literary leanings[12] underpin his project to use his prehistoric novels as a kind of experimental "demonstration" of early mankind's development of morality and social behaviour, as suggested by Darwin in his second most important work, *The Descent of Man* (1871).[13]

Darwinian ideas are also, of course, apparent in Rosny's use of the recurring motif of on-going inter-species struggle throughout the course of history. This is clearly apparent in *Les Xipéhuz*, in respect of the overt struggle that takes place between the human tribes and their stone adversaries. The conflict, caused by the tribes' need to defend their territory against the encroachment of the expanding Xipéhuz population, culminates in a pitched battle from which the humans emerge victorious, leading Bakhoûn to conclude that: "La terre appartient aux Hommes" (p. 652). Pride does not, however, obscure his obvious humanity, as he mourns the passing of the Xipéhuz, wondering "quelle Fatalité a voulu que la splendeur de la

---

[10] Linda L. Clark, *Social Darwinism in France* (Alabama: University of Alabama Press, 1984), p. 112.

[11] Rosny, *La Charpente*, pp. 21-22.

[12] J.-H. Rosny was a signatory of the infamous "Manifeste des Cinq", an open letter published in *Le Figaro* of 18 August 1887, which highlighted the perceived failures of literary naturalism in Émile Zola's *La Terre.* Regretting his participation in this venture, Rosny was later reconciled with Zola, though his ambivalent relationship with literary naturalism would be explored in greater depth in the satirical novel *Le Termite* of 1890. See Catherine Dousteyssier-Khoze, "L'Angoisse de l'influence naturaliste: *Tous Quatre* de Paul Margueritte et *Le Termite* de J.-H. Rosny", *Nineteenth-Century French Studies*, 31 (2003), 123-37.

[13] Mélanie Bulliard, *L'Enjeu des origines* (Lausanne: Archipel, 2001).

Vie soit souillée par les Ténèbres du Meurtre!" (p. 652). This sense of solidarity with other, non-human life forms signals the emergence of a kind of ecological morality in Rosny's thought, to which we shall return in the latter sections of this chapter.

The awareness that the rise to supremacy of one species is typically purchased by the extinction of another, as shown by Bakhoûn, meanwhile constitutes a further recurrent theme in Rosny's writing. In *Les Navigateurs de l'infini*, for instance, it is only the courage and technological know-how of a group of human explorers that saves the Martian Tripèdes, a species of sympathetic, intelligent beings who have fallen prey to a kind of endemic apathy, a chronic "défaut d'initiative",[14] from being overcome by enemy species collectively known as the Zoomorphes. Having explained that "[t]ous les vivants ont leur fin du monde!" (p. 61), the chief of the Tripèdes extends his rationale to his human visitors, whose superiority he has already acknowledged, in his prescience of an era in which "le déclin des hommes aura commencé" (pp. 80-81).

Contemporary anxieties about degeneration and racial decline may well have had an important influence on Rosny's ultimately pessimistic prognosis for the human race, "dont il évoque la déchéance biologique, et ceci avec une imperceptible délectation au parfum 'fin-de-siècle'", as Daniel Couegnas notes.[15] A more profound awareness of the workings of nature may, however, be discerned in Rosny's thought, inasmuch as he considers that "[l]a fin des hommes est programmée depuis leur apparition, de même que chaque être humain porte en lui la trace de l'évolution".[16] The built-in obsolescence which Rosny's Martian chief identifies as an essential characteristic of all organic life therefore serves to displace mankind from its presumed position of ultimate superiority over all other species by contextualizing humanity's rise and correspondingly inevitable fall within a broader evolutionary movement taking place over eons of geological time. This movement may be understood not as a teleological progression towards the establishment of man as nature's highest achieve-

---

[14] J.-H. Rosny Aîné, *Les Navigateurs de l'infini. La Mort de la Terre* (Paris: Denoël, 1983), pp. 7-93, p. 91. Subsequent page references to this text will be given in brackets after the relevant quotation.

[15] Daniel Couegnas, "Préhistoire et récit 'préhistorique' chez Rosny et Wells", *Europe*, no. 681-682, jan-fév 1986, 18-29, p. 26.

[16] Compère, "La Fin des hommes", p. 31.

ment, but rather as an endless succession in which the mantle of dominance passes naturally from one organic order to another. Rosny's conscious rejection of anthropocentric evolutionary perspectives is developed most explicitly in *La Mort de la Terre* of 1908, in which mankind's final destiny is recounted.

La Mort de la Terre is set several hundred centuries in the future, and depicts the decline and final disappearance of the human race after many thousands of years of intense seismic activity, during which most of the world's water resources have drained into inaccessible subterranean regions. The earth's retention of water has led to a shift in the balance of power, for, as the author explains, "les secousses sismiques [...] avaient brisé la puissance humaine",[17] through a chain of consequences beginning with drought and the ensuing dehydration to which humans and their livestock are subject, compounded by the inevitable failure of agriculture. Darwin explains that the struggle for existence, which is the basis of his theory of natural selection, need not assume the form of open conflict between living organisms: "Two canine animals in a time of dearth, may be truly said to struggle with each other which get food and live. But a plant on the edge of a desert is said to struggle for life against the drought, though more properly it should be said to be dependent on the moisture."[18] In similar fashion, Rosny's embattled humans find themselves locked in a struggle for existence against the constantly worsening drought, which has already led to the progressive reduction of the human population over the course of some five hundred centuries from around twenty-three billion to just a few thousand at the start of the text.

Demographic concerns are explored in some detail, when, adding a measure of somewhat incongruous romantic interest to the piece, Rosny creates a strong attraction between his hero, Targ, and an attractive blonde earthquake-survivor from a neighbouring tribe, named Erê. It is only by distinguishing himself as one of the community's fittest members by locating an underground water source that Targ earns the right to marry Erê and have two children, "le mariage [étant] un privilège réservé aux plus aptes" (p. 126). Yet while the restriction and / or prohibition of marriage and procreation is a draco-

[17] J.-H. Rosny Aîné, *La Mort de la Terre* (Paris: Denoël, 1983), p. 123. Subsequent page references to this text will be given in brackets after the relevant quotation.
[18] Charles Darwin, *The Origin of Species*, 2nd ed. (Oxford: Oxford University Press, 1996), p. 53.

nian step that would infringe the civil rights of the individual in any developed society, the full extent of the measures put in place as a response to further seismic activity and the consequent loss of reservoirs is even more shocking:

> À mesure que s'épuisaient les provisions, chaque oasis désignait les habitants qui devaient périr. On sacrifia d'abord les vieillards, puis les enfants, sauf un petit nombre qui furent réservés dans l'hypothèse d'un revirement possible de la planète, puis tous ceux dont la structure était vicieuse ou chétive. L'euthanasie était d'une extrême douceur. (p. 172)

Implementing the most extreme eugenic measures, including not only a strict and near complete prohibition on procreation, but also deliberate population reduction by means of compulsory euthanasia, Rosny's embattled communities seek to prolong the existence of their kind by restricting their numbers in order to limit consumption of the ever scarcer resources. Yet while the misery of their condition pushes many to end their lives before they have been required by the law to do so, environmental conditions deteriorate even more rapidly. Finally, Targ, finding himself to be the last human being alive, disdains euthanasia and sets off instead to offer himself up as prey to other life forms, better adapted than he to the new conditions of planet earth. Comprising a full development of the eschatological themes evoked in *Les Xipéhuz* and *Les Navigateurs de l'infini*, *La Mort de la Terre* represents the logical progression of Rosny's thought. The portents of mankind's final disappearance may be seen to emerge in the text through evolutionary narratives of the earth that are not, however, exclusively Darwinian in character.

Lamarck and Rosny

Although Darwin's theories exercised a considerable influence over Rosny, they were not the only evolutionary model to inform his ideas, for, as Roger Bozzetto points out: "Pour Rosny, l''évangile évolutionniste' de Lamarck est le cadre et le moteur de l'histoire de la biosphère."[19] Jean-Baptiste Lamarck's *transformiste* model of evolution,

[19] Roger Bozzetto, "Wells et Rosny, le sens d'un parallèle, les formes d'un duo", *Europe*, no 681-682, jan-fév 1986, 3-11, p. 8.

set out in *Philosophie Zoologique* of 1809, enjoyed something of renaissance in fin-de-siècle France, offering a more harmonious and progressive vision of the development of life on earth than Darwin's ostensibly conflictual theories. Asserting that environmental influences could effect direct change in an organism, and that such acquired changes could, furthermore, be bequeathed to the organism's offspring, Lamarckian *transformisme* posited the progressive development over the course of generations of ever more organically sophisticated creatures who would be increasingly closely attuned to the demands of their environment. While in the Darwinian evolutionary model the environment acts as a kind of filter, serving to eliminate the least well-adapted organisms, thereby improving future breeding stock, Lamarck's alternative system envisages living creatures engaged in a closer and more responsive partnership with their native milieu. Classifying Lamarckian thought "plus 'écologiste'"[20] than that of Darwin, Bozzetto highlights Rosny's attraction to this idea of a close cause-and-effect relationship between milieu and organism in his works. Viewed within the framework of just such a Lamarckian evolutionary narrative, the character and constitution of the inhabitants, human or otherwise, of Rosny's worlds may be read from the terrestrial "texts" in which the action takes place.

In *La Mort de la Terre*, the "extraordinaires remaniements du sol" (p. 136) that have become the norm gradually lead to the total disappearance of organic life forms, such as we know them. Mankind's imminent and inevitable disaster is clearly foretold in the drastic remodelling of the earth's surface, notably in the disappearance of rivers and, indeed, whole seas, as well as of all kinds of naturally irrigated land, including "les bois, les landes, les marais, les steppes [et] les jachères", which have been replaced by "les monts [qui] pullulaient, immenses et funèbres" (p. 138). Beyond the crippling population reductions which are the most striking effect of the geological phenomena etched so unmistakably into the earth's outer crust, the effects of the on-going catastrophe are most clearly legible in the demeanour of the earth's remaining organic population. Rosny's descriptive use of mineralogical terms in relation to the tribesmen, who variously resemble "un bloc de basalte" (p. 125), or who have "des cheveux aussi noirs que l'anthracite" (p. 122), and to the birds of

[20] Bozzetto, "Wells et Rosny", p. 8.

prey with their "immenses ailes argentines, glacées d'améthyste" (p. 123), seems almost to suggest their direct emergence from the harsh rocky surfaces of their native environment.

The effects of increasingly hostile environmental conditions are not only, however, apparent in the physical constitution of the survivors, but are also seen to be instrumental in influencing their collective psychology. Experiencing, in common with the Martian Tripèdes of *Les Navigateurs de l'infini*, a gradual loss of the will to live and a growing sense of apathy in relation to their impending end, the embattled Earth-dwellers exist in "un état de résignation douce, triste et très passive. L'esprit de création s'est éteint…" (p. 140). The generalized indifference with which the majority of the human population comes to be afflicted is, moreover, accompanied by the espousal of the kind of paganistic beliefs characteristic of early human civilizations. As Rosny explains:

> Une sorte de religion est née, sans culte, sans rites: la crainte et le respect du minéral. Les Derniers Hommes attribuent à la planète une volonté lente et irrésistible. D'abord favorable aux règnes qui naissent d'elle, la terre leur laisse prendre une grande puissance. L'heure mystérieuse où elle les condamne est aussi celle où elle favorise des règnes nouveaux. (p. 140)

This reversion to irrationalism represents a clear retrograde step for an organism characterized by its capacity to reason which is every bit as significant in relation to the generalized decline of the human species as the acute demographic shrinkage noted above. Yet while humanity is shown to be approaching the end of its reign on earth, an alternative species, even more closely related to the host environment than the remaining human and avian populations, is clearly in its ascendancy. It is in relation to the rise to power of humanity's successors that the Darwinian and Lamarckian strands of evolutionary thought evoked by Rosny converge most clearly. The alternative narrative of the earth which emerges may, in the light of modern ecological thought, be read as a clear warning about the nature of our bequest to future generations.

Towards Ecological Awareness

In *La Mort de la Terre*, the human protagonists find themselves engaged in a Darwinian struggle for existence against unrelentingly difficult environmental conditions, as we have already noted. In common with the protagonists of *Les Xipéhuz* and *Les Navigateurs de l'infini*, they must also engage in inter-species conflict with hostile beings who seek to encroach on their territory and, in respect of the Zoomorphes in *Les Navigateurs* at least, to feed upon all or parts of the human body. For the predators of *La Mort de la Terre*, the "ferromagnétaux", are strange creatures composed entirely of malleable iron, moving slowly in packs, and feeding on human haemoglobin, drawn magnetically through the surface of the skin until death ensues. Non-organic in their constitution, they resemble the Xipéhuz in respect of the fact that they are mineral in nature. More closely related, therefore, to the terrestrial environment from which they have effectively been extracted than their human adversaries could ever be, they may also be compared to the Zoomorphes, the predator species in *Les Navigateurs de l'infini*, "[qui] semblent jaillir de la terre" (p. 84), and give the impression of remaining "attachés au sol" (p. 83).

The Lamarckian theme of a cause-and-effect relation between milieu and organism thus finds its most explicit expression in Rosny's conceptualization of these exotic creatures as the very living embodiment of their native soil (Mars rather the Earth in the case of the Zoomorphes). Just as the seismic turmoil afflicting the Earth has, for instance, opened vast fissures through which the water resources vital to the survival to all organic species are being swallowed up, so the ferromagnétaux literally consume the life-blood of their human adversaries, as Targ discovers in his first encounter with them, exclaiming: "Les ferromagnétaux […] *Ils boivent ma vie!*" (author's emphasis, p. 164). Their acute hostility towards mankind may even, in Bozzetto's view, be an indication of Rosny's early intuition, "avant la percée de la modernité écologique" of certain environmentalist perspectives on contemporary humanity's troubled relationship with the Earth.[21]

[21] Bozzetto, "Wells et Rosny", p. 10.

For the ferromagnétaux exist as the embodiment of human industrial society's responsibility in relation to ecological destruction, as Rosny explains in the following:

> On commença à percevoir l'existence du règne ferromagnétique au déclin de l'âge radioactif. C'étaient de bizarres taches violettes sur les *fer humains*, c'est-à-dire sur les fers et les composés des fers qui ont été modifiés par l'usage industriel. Le phénomène n'apparut que sur des produits qui avaient maintes fois resservi: jamais l'on ne découvrit de taches ferromagnétiques sur des *fers sauvages*. Le nouveau règne n'a donc pu naître que grâce au milieu humain. (author's emphasis, p. 141)

Human industrialization is thus explicitly blamed for the creation of these sinister beings, which, although initially manageable in their restricted numbers, gradually become a force to be reckoned with. Composed entirely of mineral resources repeatedly corrupted by industrial use, the ferromagnétaux firstly defy human mastery, "ne se prêt[ant] à aucune combinaison ni à aucun travail oriénté" (p. 143), then reveal themselves as the chief living adversaries of the human species. They are therefore the incarnation of planet earth's revenge on mankind, condemned for the recklessness of its ways, as Rosny, evoking "une forme d'animisme"[22] similar to that observed in his prehistoric novels, explains: "Après trente mille ans de lutte, nos ancêtres comprirent que le minéral, vaincu pendant des millions d'années par la plante et la bête, prenait une revanche définitive" (p. 139). This revenge is shown to be decisive, as Targ, the last remaining member of the human species, disdains suicide, preferring to surrender to the predators who are the new masters of the Earth: "Ensuite, humblement, quelques parcelles de la dernière vie humaine entrèrent dans la Vie Nouvelle" (p. 220).

These closing lines acknowledge the necessary cyclicity of endings and beginnings in the extinction of mankind and the simultaneous rise to supremacy of new species, recalling, with a certain wry optimism, one of the recurring themes of literary naturalism, with life springing ceaselessly forth out of and death and decay. Rosny's relative lack of sentimentality over the final disappearance of the human race may also, however, indicate his intuition of a viewpoint nowadays known as biocentric inhumanism, associated with the Deep Ecology school of thought. Greg Garrard defines "mainstream envi-

---

[22] Bulliard, *L'enjeu des origines*, p. 46.

ronmentalism" as the range of ecological ideas espoused by "people who are concerned about environmental issues such as global warming and pollution, but who wish to maintain or improve their standard of living, […] and who would not welcome radical social change".[23] While its advocates are now sufficiently influential to lobby politicians and captains of industry, opponents argue that the real impact of their views can only ever be restricted because of their firmly anthropocentric outlook, always placing human needs before those of the environment and other species.

Deep ecological perspectives, on the other hand, adopt a clearly biocentric perspective as their starting point, acknowledging that "[t]he well-being and flourishing of human and non-human life on Earth have value in themselves", and that "[t]he flourishing of human life and cultures is compatible with a substantially reduced human population. The flourishing of non-human life *requires* a smaller human population".[24] Rosny's unsentimental depiction of a situation in which the very survival of the human race is firstly considered to be linked to effective population limitation, and in which other species subsequently grow in dominance as humanity retreats, suggests that he may well have anticipated the principles of biocentric inhumanism in his own ecological reflections. His assertion that human industrial society's direct responsibility for the creation of the ferromagnétaux "a beaucoup préoccupé nos aïeux" (p. 141) meanwhile hints at the issue of the ecological guilt of his own contemporaries, and this in an era prior to the foundation of anything resembling an environmentalist literary canon.[25]

It is in this connection, in effectively anticipating the problems that man's abuses of the Earth may be storing up for the future of his own descendants, that Rosny's writing may be seen to be truly innovative. His multi-layered conception of a terrestrial text has a clearly natural aspect, variously revealed in the forms of the paleontological treasure-trove beneath our feet, in the geological phenomena that have been continuously on-going since the Earth's beginnings,

---

[23] Greg Garrard, *Ecocriticism* (London: Routledge, 2004), p. 18.

[24] Garrard, *Ecocriticism*, p. 21.

[25] Romain Gary's *Les Racines du ciel* of 1956, recounting the story of a fictitious campaign to save the African elephant from extinction, is widely recognized as the first "environmentalist" French-language novel, in the modern sense.

and in the advancement of evolutionary narratives through the successive reigns of different life forms from the very origins of life to the advanced organisms of an almost unimaginably distant future. To these natural realizations of the "terrestrial" text, however, there corresponds a clearly manufactured, *human* aspect, as, for example, in the archaeological discovery of Bakhoûn's record of mankind's historic victory over the Xipéhuz in the *livre lapidaire* which accompanies the mineral relics of the now extinct species. Similarly, the "annales anciennes" of the "écrivains de l'âge radioactif" evoked in *La Mort de la Terre* (pp. 142-43) serve as a written warning for future generations of humans about their impending doom, effectively mirroring that which may be read in the transformation of the Earth's crust as the age of geological catastrophe begins.

It is, however, a highly ironic awareness of man's relationship with his environment that Rosny evokes in titling the text in which he deals most explicitly with the issue "*La Mort de la Terre*". For while the graceful surrender of the heroic Targ graphically signals the death of humanity, the Earth actually persists in all its grandeur, with the newly dominant species of ferromagnétaux living in apparently closer harmony with their surroundings than mankind ever did. As Targ reflects:

> Il [l'homme] fut le destructeur prodigieux de la vie. Les forêts moururent et leurs hôtes sans nombre, toute bête fut exterminée ou avilie. Et il y eut un temps ou les énergies subtiles et les minéraux obscurs semblèrent eux-mêmes esclaves…
> - Cette frénesie même annonçait la mort de la terre […], la mort de la terre pour *notre* Règne! (p. 219)

Explicitly recognizing the arrogance of the anthropocentric perspective that has led him, along with the rest of his species, to historically articulate the death of mankind as the end of the world itself, Targ finally understands the insignificance of mankind's existence in relation to the immensity and permanence of the Earth. Yet while a part of Targ's history and that of all his collected forebears and non-human successors may, as we have already suggested, be read from the Earth, an alternative and complementary account of his struggle is recorded in the written account of events which constitutes *La Mort de la Terre* itself. Simultaneously revealing what we may term the "récit raconté" of the Earth and the "récit racontant" of the written word,[26] Rosny seems to compare the permanence of the terrestrial text, in its various mineral and literary forms, to the ephemeral character of our own physical existence. In so doing, he issues an ominous warning which appears clearly foresighted in the light of growing ecological concerns about humanity's relationship with the environment and our own final destiny.

[26] Maraud, "Le Texte à l'origine de l'histoire", p. 105.

André Gide, Eugène Rouart and *le retour à la terre*

David H. Walker

Abstract: *The Third Republic not only revolutionized France's education system, but also sought to radically reform land-use practices. To this end, agronomists such as André Gide's friend and correspondent, the future Sénateur Eugène Rouart pioneered agricultural education and engaged in small-scale agricultural research across France. This chapter considers the extent to which their reformist zeal contrasted with the chauvinistic blood-and-soil regionalism of Maurice Barrès, as well as the importance of agricultural metaphor in Gide's writing.*

In 1897 André Gide famously published *Les Nourritures terrestres*, a book whose title indicates eloquently enough its aim of encouraging the reader to celebrate earthly delights. In the Preface for a re-edition in 1927 Gide explains that his purpose had been to bring literature back from the abstractions of the fin-de-siècle – symbolism, decadentism, and the like: "la littérature sentait furieusement le factice et le renfermé … il me paraissait urgent de la faire à nouveau toucher terre".[1] The work consists of eight "Livres" and an "Envoi", and the significance of this structure is underlined via the Gidean device of the *mise en abyme*, whereby at the heart of the volume is reproduced an image in miniature of the shape and substance of the volume as a whole. This *mise en abyme* is a section entitled "La Ferme", in which the narrator successively opens eight doors and takes us on a tour of

[1] "Postface" to a re-edition in 1927, in André Gide, *Romans, récits et soties, œuvres lyriques* (Paris: Gallimard, Bibliothèque de la Pléiade, 1958), p. 249. Hereafter referred to in the text as RRS.

the various parts of the Farm: the barns (les granges), the corn-loft (les greniers), the dairy (la laiterie), the cowshed or shippon (l'étable), the fruit store (le fruitier), the cider press (le pressoir), the still-room (la distillerie), the toolshed (les remises) – or more precisely the shed where equipment such as ploughshares and carts are kept. This last place is the culmination of an increasingly intense – and systematic – rumination on the products of the earth: from seeds, through the growth cycle to the concentrated alcoholic essence of nature. So in the "remise" the narrator becomes aware that it is time for new departures, in search of "Vous toutes, possibilités oisives de nos êtres, en souffrance, attendant" and hitches up a cart to set off for pastures new: "La dernière porte ouvrait sur la plaine" (RRS, p. 214), we learn. This interplay, or oscillation, between the sedentary agricultural life and that of the nomad constantly in search of new pastures will be crucial for Gide. Though this is not the place to go into its implications, it suffices to confirm the extent to which Gide uses "histoires de la terre" as allegories for human activity and human attitudes. One could point out also that his pioneering "sotie" of 1895, entitled *Paludes*, depicting a writer's attempt to extricate himself from the miasma of Parisian intellectual life, also adopts an agricultural metaphor for the process: the "hero" signals his somewhat risible ability to retrieve something from his plight by moving on to a book entitled *Polders* (RRS, p. 146).

If farming is at the heart of Gide's early writing in this way, it is partly because farming was an empirical reality for Gide while he was embarking on his literary career. In February 1893 he made the acquaintance of Eugène Rouart. The friendship quickly blossomed: they shared many interests, literary, musical and artistic; both were homosexual; and a distinctive personal chemistry formed a curious basis for a relationship that had its ups and downs but was to last right through to Rouart's death in 1936.[2] Eugène was the son of Henri Rouart, an industrialist, engineer, impressionist painter, and friend of Degas. To escape from the overbearing dominance of this massively successful father, Eugène decided to go into agriculture, and in August 1893 sat the exams for entry to l'École d'Agriculture de Grignon.

Founded in 1829 on an estate near Versailles, occupying a chateau built by Louis XIII, Grignon was to be one of some twenty

[2] See André Gide - Eugène Rouart, *Correspondance* I, 1893-1901; II, 1902-1936 (Lyons: Presses universitaires de Lyon, 2006). Henceforth indicated in the text by GR, with volume and page number.

such establishments which by the 1880s were helping the Third Republic revolutionize French agriculture through developing the technology of farming and the sciences of crop development and animal husbandry. If Jules Ferry's schools produced the famous "Hussards noirs de la république", it is equally true to say that under the stewardship of Jules Méline, the first man to preside over a separate Ministry of Agriculture, the *Ecoles d'agriculture* like Grignon sent out across the country an analogous army of pioneering "ingénieurs agronomes" bent on modernizing the countryside. For Méline the watchword was *Le Retour à la terre* to offset the effects of excessive industrial production, as he argued in his book of 1905; after World War I he would raise the stakes: *Le Salut par la terre.*[3] As for the rôle played by the schools of agriculture, one need only read Georges Duby and Armand Wallon's *Histoire de la France rurale*[4] to realize that Eugène Rouart's professors were those to whom historians look back now as trailblazers for modern farming. Names little known outside this specialist sphere made their mark within it in no uncertain terms. André Sanson, an expert in zootechnie and animal genetics, as well as being a revered professor, was witness at Eugène's wedding; the work of another professor Daniel Zolla on cattle breeding, is drawn on by Duby. Lucien Brétignière, a pioneer in plant biology and close associate of the Vilmorin family, expert in scientific fertilizers and silage, was a *condisciple* and ultimately wrote Eugène's obituary.[5] A key mentor of Eugène, François Berthault, was responsible for reconstituting the vineyards of the south-west after the phylloxera disaster of the late nineteenth century – and wrote one of the books telling how the salt plain of L'Habra, in the Sahel, had been turned into fertile farming land.[6] The pioneering enthusiasm of these authorities communicated itself to Rouart and his fellow-students.

Eugène's results at the *concours d'entrée* did not secure him immediate entry to Grignon, and to his initial dismay he had to do a *stage* at the *Ecole d'Agriculture* in Montpellier – a long way from his

---

[3] See Jules Méline, *Le Retour à la Terre* (Paris: Hachette, 1905); *Le Salut par la terre*, (Paris: Hachette, 1919).

[4] *Histoire de la France rurale*, Vol. III, "Apogée et crise de la civilisation 1789-1914", by Georges Duby, Armand Wallon, Maurice Agulhon, Gabriel Désert, Robert Specklin (Paris: Seuil, 1992).

[5] *Bulletin des Anciens Élèves de Grignon*, no. 8, août 1936, pp. 262-66.

[6] See *Correspondance* Gide-Rouart, II, p. 198.

Paris home. Gide offered him encouragement, pointing out that Montpellier was a prestigious centre for the teaching of Viticulture; and that his uncle Charles Gide, at that time a professor at the Faculty of Economics at the University of Montpellier, would be happy to welcome Rouart to his home (GR, I, p. 117). In fact during the winter of 1893 Rouart attended Charles Gide's lectures on "la solution sociale" (GR, II, p. 498).

In the meantime, André Gide had embarked on his life-changing trip to Algeria, where as is well-known he fell ill, came close to death, and recovered with a new-found sense of the joys of existence, determined to throw off the shackles of religion and conventional morality in order to live a fuller life. Throughout his lyrical convalescence, however, he is pursued by letters from Eugène who is learning to tell a horse's age from examining its teeth, and who recounts how he has groomed a cow: "– j'en repanserai une la semaine prochaine, dégoûtant métier" (GR, I, p. 144, letter of 25 January 1894). In April 1894 Rouart finally enrolled at Grignon and continued his studies as Gide, after a brief return to Paris, convalesced in Switzerland. In one of his letters addressed to Grignon, Gide points out that he has gone up significantly in the estimation of the locals in Switzerland since they have learned he is in correspondence with an "agriculteur" (GR, I, p. 225). Soon, however, Gide is back in Algeria, where in January he has a momentous encounter with Oscar Wilde and confirms his homosexuality.

But at the same time he is receiving letters from Rouart about his agricultural studies: notably on 19 February:

> Puisque tu es à Biskra centre du Dattier, dans tes moments perdus de promenade ne pourrais-tu me prendre quelques renseignements sérieux sur cette culture; et me trouver des photographies de champs de dattiers, et aussi représentant les principales opérations culturales; comme fécondation, récolte, etc.; aussi le plan d'une datterie. (GR, I, p. 251)

Clearly Rouart needs help researching an essay he has to write. Gide conscientiously carries out the task; and just as the request for information may have a familiar ring to all academics, so Gide's response, dated 10 March, is not without its contemporary pertinence:

> Je me suis adressé à diverses personnes pas trop incompétentes; en particulier à Monsieur Fau – qui m'a renvoyé à Monsieur Colombo qui m'a renvoyé à des livres: un surtout écrit par quelqu'un précisément de Grignon. J'ai pu me convaincre qu'à

moins de me livrer à un énorme travail je ne pouvais t'envoyer rien de sérieux sur la culture du palmier. La compagnie … qui s'en occupe sérieusement en est encore à la période empirique et n'a pu faire d'assez longues expériences encore pour comprendre pourquoi les cultures rataient sitôt qu'on ne suivait plus la routine arabe. On a dû arriver à cette mode (les Arabes) par sélection naturelle, par exhaustion; mais rien ne prouve encore que cette méthode est la meilleure ou la seule bonne, car les Arabes ne pouvaient disposer de toutes les ressources que la science et l'argent et les outils français apportent. Quant à cette méthode elle-même tu le trouveras longuement exposée dans plusieurs livres – que j'ai (quelques-uns) et qui m'ont en partie assez intéressé. (GR, I, p. 255)

So Gide is not merely an observer of agricultural method; he is actively pursuing fieldwork. Very soon, this fieldwork will become an urgent matter for him. In May 1895 Gide's mother died and he became the owner of an estate in Normandy, notably of the farm of La Roque-Baignard, which had been a source of worry every since Mme Gide mère had reluctantly taken it on on her marriage. She had written to her son in 1894:

Combien j'avais eu raison de faire des difficultés pour prendre La Roque, propriété de plus extrêmement difficile à vendre, vu la dépréciation des propriétés depuis si, si longtemps – ou à vendre à vil prix.[7]

What Gide no doubt discovered is what historians have subsequently underlined: from the 1860s investment in agricultural equipment and infrastructure to improve yields had become much less attractive a prospect than investment in the industrial and banking sector or in state bonds. "*Exploitants*" at all levels, from gentleman farmers to peasant smallholders, found it more advantageous to invest profits in the stock market rather than, as Gabriel Désert puts it, "d'augmenter son capital d'exploitation et de roulement, clé de tout progrès technique".[8] The "rentier du sol", following this trend, was soon being accused of "détournement de capitaux au détriment de l'agriculture".[9] Though it was still possible to make a living from the land, by virtue of reasonable prices generated by favourable market conditions, "Les paysans cueillaient les roses, ils oubliaient les epi-

[7] André Gide, *Correspondance avec sa mère,* ed. by Claude Martin (Paris: Gallimard, 1988), p. 417.
[8] *Histoire de la France rurale.* Vol. III , p. 236.
[9] *Histoire de la France rurale.* Vol. III , p. 237.

nes."[10] The long-term deleterious effects of this lack of investment would soon be widespread. And when market prices began to fall, the relationship between owners and the tenant- farmers who worked the land for them became extremely fraught. As Désert explains, whenever leases came up for renewal, tough negotiations were the order of the day: "La lutte [...] aboutit dans quelques régions à de véritables grèves organisées au moment du renouvellement des baux, aspect original de la lutte des classes dans un monde qui la connaît peu."[11] The flight from the land having reduced the number of prospective farmers, those remaining could increasingly impose whatever conditions they liked: "Les propriétaires sont dans l'obligation, pour trouver des fermiers, de consentir des réductions sur les baux ou même, afin de conserver ceux qui sont en place, de leur accorder des remises." Some were forced to agree to substitute "le fermage en argent au fermage en nature" and: "Certains, faute de trouver des candidats, sont obligés de reprendre en main l'exploitation de leurs terres."[12] So Gide found himself on the uncomfortable side of the "grande déroute du rentier du sol":[13] he had an estate which had suffered from under-investment, and a population of farmers intent on extracting maximum advantage from the sellers' market in which they found themselves.

Now at this very moment Rouart was impatient to become actively involved in a real agricultural concern. He had already come close, in January 1895, to throwing up his studies, and taking on an actual farm (GR, I, p. 248). So that clearly, when Gide turned to him for advice, he was eager to lend a hand. To begin with, he recommended to Gide a fellow student from Grignon, Lecomte, who did indeed come and do some work for him at La Roque in the summer of 1895. Rouart congratulated Gide on having taken the young man on and for giving him a chance to show off his skills (GR, I, p. 293, 19 August 1895). However the following year it had became clear to Gide that La Roque could well ruin him and he asked Rouart to come and spend a few days with him there, to study the situation on the estate and propose some remedies as well as offer some technical advice

[10] *Histoire de la France rurale*. Vol. III , p. 237.
[11] *Histoire de la France rurale*. Vol. III , pp. 374-75.
[12] *Histoire de la France rurale*. Vol. III , p. 375.
[13] *Histoire de la France rurale*. Vol. III , p. 375.

on agricultural matters.[14] This proved to be the start of an adventure which would run and run, and would inspire an important part of Gide's novel *L'Immoraliste*.[15]

After a fact-finding visit in August 1896 Rouart proposed to Gide that he should come back to La Roque with one Marie-Émile-Ernest Claudel, an alumnus of l'École nationale d'Agriculture de Grignon, a graduate of 1887, who by 1896 had become a member of the teaching staff: "répétiteur-préparateur d'agriculture et d'économie". Eventually a date was agreed for the visit of this "grignonnais très calé".[16] Now, whereas Michel in *L'Immoraliste* is confronted by Charles, a young man of 17, just finished a placement on a model farm, for whom he forms an attachment of a nature he is reluctant to acknowledge, it should be pointed out that Claudel had graduated nine years previously and was aged around forty at the time. So what *L'Immoraliste* owes to Claudel therefore is a strictly technical perspective on the management of an agricultural enterprise. Rouart himself, fresh out of Grignon in 1897, worked his placement on a model farm in the summer of 1897 – and from there he was to accompany Claudel on an inspection visit to La Roque in July 1897.[17] As for the attraction that in the novel, the young Charles exerts on Michel, to the extent of actually clouding the latter's better judgement in respect of his impetuous ideas on farm management, doubtless it was Rouart's ill-considered enthusiasm, as well as his contagious personal charm, which contributed to these other aspects of the novel's plot. On reading Gide's novel in June 1902, Rouart, still full of advice on how best to run La Roque, would nonetheless protest: "je ne fais en rien mon petit Charles" (GR, II, p. 116).

Following the publication of *L'Immoraliste*, Gide was to recall in a letter of 27 November 1902 what Rouart and Claudel between them had contributed to the situation of one farm in particular, the rent

[14] Letter to Paul Valéry, 19 May 1896, in André Gide - Paul Valéry, *Correspondance,* ed. by Robert Mallet (Paris: Gallimard, 1955), p. 266.

[15] See Jacques Copeau, *Journal 1901-1948* (Paris: Éditions Seghers, 1991); repr. by Éditions Claire Paulhan, 2000, vol. 1, 1901-1915, p. 199, 31 mai 1905: "Les sources de *L'Immoraliste*: La Roque – exploitation d'une ferme sur le conseil de Rouart".

[16] Letter to Valéry, 14 September 1896, in André Gide - Paul Valéry, *Correspondance,* p. 276.

[17] André Gide, *Journal 1887-1925,* ed. by Éric Marty (Paris: Gallimard, Bibliothèque de la Pléiade, 1996), p. 263.

for which had been set by Mme Gide *mère* at too low a level and which, as a result, the tenant farmer could pay without effort,

> […] avec si peu de peine qu'il négligeait de cultiver la totalité de la terre, mais laissait se prendre de chardons, de genêts et de joncs une partie. C'est beaucoup pour remettre en état cette partie négligée, que sur les conseils de Claudel et les tiens j'ai repris la ferme à mon compte. (GR, II, pp. 125-26)

In this respect Gide was resorting to the very expedient that many others were being reduced to at the time, as has been pointed out above. He recalls the advice of Rouart and Claudel: "Vous pensiez tous deux que, bien cultivée, cette ferme devait pouvoir rapporter beaucoup plus" (GR, II, p. 126); these words invite comparison with what young Charles says to Michel in *L'Immoraliste*:

> Charles ne me dissimulait point l'irritation que lui causait la vue de certains champs mal cultivés, d'espaces pris de genêts, de chardons, d'herbes sûres ; il sut me faire partager cette haine pour la jachère et rêver avec lui de cultures mieux ordonnées. (RRS, p. 414)

Michel asks why it matters that land is left fallow, so long as the farmer pays his rent, and is told:

> Ne considérant que le revenu, vous ne voulez pas remarquer que le capital se détériore. Vos terres, à être imparfaitement cultivées, perdent lentement leur valeur. (RRS, p. 414)

So, like Michel, Gide took over the running of these farms which the tenants had allowed to run down.

> J'ai mis tout mon amour et tout mon amour-propre à mettre en excellent état terres et bâtiments, et si cela m'a coûté assez cher, du moins ai-je la conviction que la ferme a beaucoup gagné; la conviction aussi que si malgré tout la ferme ne me rapporte pas actuellement tout ce que vous m'en promettiez, la faute en est à l'absence de surveillance personnelle et à des tas de … complaisances que je suis trop loin pour empêcher. (GR, II, p. 126)

Here we can see clearly that in taking over the management of the farm himself, the system Gide had put in place on the advice of Rouart and Claudel was still costing him dear. It is true that Rouart had warned him that he would have to invest extensively in the "capital de roulement" before being in a position to reap any dividends:

"Songe que tu vas probablement être obligé de mettre de l'argent à La Roque si tu veux que ça marche bien" (GR, I, p. 397, 27 May 1897). However, the extent of Gide's dissatisfaction can be seen in the apologetic letter Rouart wrote to him in March 1898:

> J'ai vu Claudel, il paraît que ça marche à La Roque: un peu de dépenses c'est vrai, mais vers quelles économies industrielles et superbes marches-tu? [...] Claudel s'intéresse beaucoup à la chose, je l'ai bien vu à plusieurs indices, et fait marcher Désauney qui grogne un peu. (GR, I, p. 460, 10 March 1898)

This Désauney, Gide's bailiff or farm manager, would subsequently turn out to have been operating some extensive embezzlement scams at Gide's expense, so in drawing attention to his role Claudel was trying to deal with the principal causes of the financial problems that La Roque was giving rise to. But Gide seemed unconvinced as to the effectiveness of the Grignon expert; in July 1898, Rouart had to promise he would speak firmly to Claudel. For his part, Rouart hangs onto the hope that his intervention has not caused more trouble than it was worth: in 1899 he would finally feel able to write: "Je suis ravi que cela aille bien à La Roque, je serai heureux de te voir accroître les revenus par ton faire valoir et ne pas t'avoir été néfaste en cela" (GR, I, p. 522, 19 May 1899).

The story continues, with more than its fair share of twists and turns; and it was only in 1905, ten years after his mother's death and Rouart's initial intervention, that Gide could finally get rid of the financial burden by selling off the estate (in the novel Michel, soon impatient with the prudence and rationality required of farm management, after having been caught poaching on his own land by Charles now grown distastefully mature and sensible, sells up much more quickly). By this time Rouart was no less relieved than the owner himself, for he had been nursing some guilt about having pushed Gide, out of a somewhat misguided enthusiasm for the latest technical and managerial methods, into the ill-considered strategy that Gide had consoled himself with denouncing in *L'Immoraliste*: "J'espère que La Roque te donnera moins d'ennuis à l'avenir – j'ai toujours le regret de t'avoir poussé à faire de la gestion directe" (GR, II, p. 223, 31 October 1905).

So in the same year that Jules Méline published his key book, *Le Retour à la terre*, Gide was actually distancing himself from the spade-work, so to speak. It was undoubtedly a relief to him to be able to do so; and the insight into the travails it cost him to be so closely associated with farming may alert us to a peculiarly personal dimension of a literary-ideological polemic Gide pursued during this same period. We began by pointing out that a dialectic between nomadism and sedentary farming informs *Les Nourritures terrestres*. However, far from equivocating on this theme, Gide reacted vigorously when Maurice Barrès, one of the most celebrated figures of the era, published his novel *Les Déracinés* several months after the appearance of *Les Nourritures terrestres*. In a blunt rebuttal of the doctrine that young people, instead of moving to Paris and other foreign parts where they become subject to deleterious influences, should remain attached to their native soil, Gide began his article "A propos des *Déracinés* de Maurice Barrès", in *L'Ermitage* of February 1898, with the words: "Né à Paris, d'un père uzétien et d'une mère normande, où voulez-vous, monsieur Barrès, que je m'enracine?".[18] This ringing challenge would continue to resound, as nationalists and reactionaries sought to dispute the horticultural data on which Gide based his ethical argument in favour of uprooting, transplanting, and *disponibilité* as the secret of robust growth in humans as well as plants. What came to be known as "La Querelle du Peuplier" rumbled on for over six years, and by 1905 Gide was arguing the point with Charles Maurras[19] after Rémy de Gourmont, Émile Faguet, and an aristocratic farmer called Le baron de Beaucorps had had their say. Eugène Rouart himself had weighed in with an article in 1903, following the republication of Gide's piece in his volume *Prétextes*; it was in debating the issues again with his farmer friend that Gide rehearsed his concluding argument (GR, II, pp. 159-68).

Although Gide had hoped to be free of his responsibilities as a gentleman farmer, the saga of La Roque would not be completely liquidated till 1909; and perhaps it is no coincidence that at this point, Gide wrote one of his most telling instances of looking to the earth primarily for pretexts for literary and intellectual undertakings. In November 1909 he published in *La Nouvelle Revue Française* an article (the last of a series of three) taking to task the reactionary outlook of a group of Nationalists who were seeking to reformulate the principles of French art and culture in terms of traditional classicism and a nar-

[18] See André Gide, *Essais critiques* (Paris: Gallimard, Bibliothèque de la Pléiade, 1999), p. 4.
[19] See Gide, *Essais critiques*, pp. 121-26.

row patriotism. What provoked Gide was that they were harking back, not just to "la terre et les morts", as Barrès put it, but to the seventeenth-century notion that "Tout a été dit", that all that is left to us is to repeat as elegantly as we can what the *Anciens* originally discovered and formulated so well. Taking up once more the agricultural metaphor Gide parodies this doctrine thus:

> O classiques grecs, latins, français! Vous avez pris les bonnes places. Le sol ingrat qui nous demeurerait en partage risquerait d'abîmer vos outils; la moisson qu'on y peut espérer ne paiera jamais notre peine; mieux vaut, reprenant de vos mains la charrue, la ramener dans le sillon profond que vous traçâtes.[20]

Gide likens this stance to the economic theory of David Ricardo, one of whose premisses was that agricultural pioneers or new settlers, the first to install themselves on the land, take possession of the most fertile parts, so that those who arrive later find only the "sol ingrat" that has been left for them to scrape a poor living from. To this pessimistic view Gide opposes the economics of Henry Charles Carey: no, he argues, it is not the most fertile land that is cultivated first, but the easiest. The truly rich lands are so fertile that they are massively occupied by forests and jungles and undergrowth, often containing swamps and marshes, which makes them impenetrable and off-putting to early settlers: "La terre la plus riche est la terreur du premier émigrant". So the first farmers leave them alone, and it is only the bolder ones who come after who tackle the challenges they present – and on venturing into these murky, overgrown, "terres peuplées d'animaux sournois et féroces; terres marécageuses, mouvantes, aux exhalaisons délétères", they discover that they are "terres inéspérément fécondes".[21]

The metaphorical value of this carries considerable force. As for the literal truth of what he is asserting, Gide is confident about it because he has it from a book published in that year by his uncle the economist Charles Gide, a notable pioneer and theorist of cooperative organizations, especially in agriculture. He acknowledges however, that the main literal application of Carey's theory has been all but exhausted by virtue of the great progress made by agronomists – and colonials: "Si meurtrières qu'elles aient été d'abord, nos généreuses

[20] See "Nationalisme et littérature", in, Gide, *Essais critiques*, pp. 195-99.
[21] Gide, "Nationalisme et littérature", p. 196.

plaines de la Mitidja et du Sahel à présent sont apprivoisées".[22] Moreover, if Gide knows this to be so, it is because he is well-acquainted with some of the men who did the work. Claudel at this point is actually farming an estate in North Africa, implementing François Berthault's research into reclaiming and cultivating the salt plains of L'Habra.[23] As for Gide, he is happy to leave the digging to others, rather more concerned, at this stage, to stress what is left to do for writers and intellectuals setting out to explore and cultivate the figurative "badlands" of the human spirit.

[22] Gide, "Nationalisme et littérature", p.197.
[23] See GR, II, pp. 197, 199: letters of November 1904.

Down to Earth: André Malraux's Political Itinerary and the Natural World

Martin Hurcombe

Abstract: *The intellectual and ideological engagements of André Malraux are read here through the figures of flight and return to earth. If flight affords a liberating perspective on the human condition in Malraux's early work, this chapter contends that the Spanish Civil War and the Second World War represent a brutal but necessary return to earth, characterized by the Resistance fighters of the* maquis *and the grim endurance of the French peasantry.*

André Malraux's decision in 1945 to become a member of Général de Gaulle's government signalled not only the beginning of an association that would last until the latter's death, but also what appeared to many a radical political transformation. One of France's foremost anti-fascists, and an apologist for revolutionary action in the pre-war period, Malraux emerged from the Second World War entranced by the mystique of de Gaulle and an apologist for nationalism. The possibility of Malraux's political itinerary from Communist fellow traveller and author of revolutionary fiction to supporter of de Gaulle, recent studies have suggested, lies in his evolving metaphysical conception of the human condition.[1] The present chapter adopts a similar approach, but locates this itinerary in Malraux's shifting depiction of the relationship between humanity and the natural world. Underlying this evolving relationship, this chapter will argue, we can discern a recon-

[1] For example, Geoffrey T. Harris, *André Malraux: A Reassessment* (London: Macmillan Press Ltd., 1996) and Gino Raymond, *André Malraux: Politics and the Temptation of Myth* (Aldershot: Avebury, 1995). Subsequent references to source texts will appear as page numbers either in the body of the text or in footnote form.

figuration of time and space that casts further light on Malraux's journey from the politics of the revolutionary left to those of the conservative right.

Initially, we will consider this relationship in Malraux's three Asiatic novels: *Les Conquérants* (1928), *La Voie royale* (1930), and *La Condition humaine* (1933). These novels, where the natural world is associated with a deathly and paralysing torpor against which the novels' characters attempt to act meaningfully, establish a challenge that is at once political and metaphysical, but also adumbrate an essential antagonism between humanity and the natural world. This antagonism persists in Malraux's anti-fascist works of the mid and late 1930s: the novels *Le Temps du mépris* (1935) and *L'Espoir* (1937), and Malraux's only film, *Sierra de Teruel* (1939). Now, however, it is challenged by the fraternal revolt of Malraux's characters. Here, we will examine the role played by flight and what Malraux himself identifies as the *retour sur terre*: the rediscovery of the world following a near-death encounter with the forces of nature and the concomitant discovery of an age-old collective human struggle to coalesce with the natural world through which humanity gains some of the latter's permanence. This chapter will conclude with an examination of the relationship between humanity and the natural world in Malraux's fiction and memoirs of the Second World War, where Malraux's association with Gaullism begins, in his final novel, *Les Noyers de l'Altenburg* (1948), and in his 1967 essay *Antimémoires*. The latter, published whilst he was a member of de Gaulle's government, fuses fiction and autobiography, reusing passages from several of Malraux's novels in order to suggest a continuity of thinking between the revolutionary and the Gaullist phases. In both works, the sense of human permanence revealed in the anti-fascist works is rediscovered in the French people encountered during the Second World War and, in particular, in the course of Malraux's own Resistance activities. Here again, this chapter will argue, the emphasis is on coalescence with the natural world rather than on the integration of the human into the natural world, as Gino Raymond has suggested.[2] More importantly, it will argue that it is the resulting reconfiguration of time and space, predicated upon a gradual reconfiguration of humanity's relationship to the

[2] Raymond, *André Malraux: Politics and the Temptation of Myth*, p. 43.

natural world and transposed onto the national context of France, that helps to explain Malraux's political itinerary in the post-war years.

Adopting Mikhail Bakhtin's terms, we will examine to what extent the natural world in the Asiatic novels constitutes a horizon against which the individual hero acts. For Bakhtin, in our every-day lives, we experience the world around us not as an all-encompassing and ultimately reassuring environment, but as a problematic horizon against which the purposive and future-orientated self is condemned, with a certain anticipation of Sartrean existentialism, to act.[3] However, it is the encounter with the other and the resulting aestheticization of human activity that transform the world from a problematic horizon to an all-encompassing environment in which human action gains aesthetic form and temporal depth. We will therefore consider to what extent flight and the return to earth can be understood as leading to the discovery of such an environment in opposition to the problematic horizon of the early novels, and how the Bakhtinian notion of environment helps to understand Malraux's concept of the nation.

The Asiatic Novels

Malraux's first three novels are set against Asia's anti-colonial struggles of the 1920s; the action of both *Les Conquérants* and *La Condition humaine* is located in China, while that of *La Voie royale* is located in the former French colony of Indochina. Here, the Asian climate is fundamental in establishing a conflict between humanity and the natural world that is crucial to Malraux's conception of the human condition more generally. This conflict resides in the natural world's apparent permanence in opposition to the Malrucian hero's sense of transience; Perken in *La Voie royale* and Garine in *Les Conquérants*, both examples of the Malrucian adventurer, act in the awareness of their own physical decline and eventual deaths, but do so in an environment marked by torpor and stasis suggestive of nature's permanence. The insects that Claude and Perken encounter in *La Voie royale* live "dans une immobilité d'éternité" that results from their

[3] Mikhail Bakhtin, "Author and Hero in Aesthetic Activity", in *Art and Answerability: Early Philosophical Essays*, ed. by Michael Holquist and Vadim Liapunov, trans. by Vadim Liapunov and Kenneth Bostrom (Austin: University of Texas Press, 1990), pp. 4-256 (p.97).

total integration "dans ces bois fumants de commencement du monde".[4] Such an environment not only denies the validity of the individual, but also seeks to draw and to dissolve the latter into itself:

> Quel acte humain, ici, avait un sens? Quelle volonté conservait sa force? Tout se ramifiait, s'amollissait, s'efforçait de s'accorder à ce monde ignoble et attirant à la fois comme le regard des idiots, et qui attaquait les nerfs avec la même puissance abjecte que ces araignées suspendues entre les branches. (p. 67)

The same torpor and stasis mark the climate of *Les Conquérants* and *La Condition humaine*. While the descriptions of the natural world in *La Voie royale* often emphasize the impenetrability of the jungle, the way in which its muddy soil clings to the adventurer, hindering his progress by weighing him down and drawing him in, *Les Conquérants* and *La Condition humaine* suggest the same stasis and resistance to action through periodic evocations of the climate in which the individual acts.[5] Thus, the news of the general strike in Canton reaches the narrator's steamer under the immobility of a leaden sky: "Jusqu'à l'horizon, l'océan Indien immobile, glacé, laqué – sans sillages. Le ciel plein de nuages fait peser sur nous une atmosphère de cabine de bains, nous entoure d'un air saturé."[6] In *La Condition humaine* nature's permanence is contrasted directly with the finitude of humanity as the condemned revolutionaries await their death, contemplating the shadows that shorten around them. Here the shortening of shadows is no symbol of what awaits the novel's heroes; there is no pathetic fallacy at work. Rather, it is suggestive of a constant, cosmic movement from which humanity is excluded:

> Les ombres se raccourcissaient peu à peu: les regarder permettait de ne pas songer aux hommes qui allaient mourir là. Elles se contractaient comme tous les jours avec leur mouvement éternel, d'une sauvage majesté aujourd'hui parce qu'ils ne le verraient plus jamais.[7]

Political action is thus juxtaposed to the permanence and indifference of the natural world; the relationship that exists between the characters

---

[4] André Malraux, *La Voie royale* (Paris: Livre de poche, 1987), pp. 66-67. First published in 1930.

[5] This motif also appears in *La Voire royale*. See pp. 48-49, for example.

[6] André Malraux, *Les Conquérants* (Paris: Livre de poche, 1995), p. 49.

[7] André Malraux, *La Condition humaine* (Paris: Folio, 1989), p. 270.

of the early novels and the natural world is that which exists between Camus's *homme absurde* and the world, more generally characterized as "cette confrontation entre l'appel humain et le silence déraisonnable du monde".[8]

This confrontational relationship also expresses itself in the hostility that Malraux's heroes discern in their relationship with the world around them. For Claude in *La Voie royale*, the forest becomes the enemy with which he must do battle as he attempts to disengage artworks from the ruins of temples that have been reclaimed by ever-encroaching nature. In the case of Perken and Garine, it is the Asian climate more generally that undermines both heroes' health.

While such a conceptualization of the natural world reflects Malraux's understanding of the human condition more generally, it also results directly from the characters' engagement in action. As Bakhtin writes, action shapes our perception of the world around us, transforming the "given makeup of the external world of objects, [breaking] up the *body* of an object's present state. The anticipation of a future actualization permeates the entire horizon of the action-performing consciousness and dissolves its stability" (p. 45). For the characters of the Asiatic novels, therefore, the world constitutes a horizon against which they act, the natural world being seen as at best an obstacle, at worst an adversary. Thus, as Claude attempts to dislodge a statue from the walls of one ruined temple, the jungle around him, the very rock of the temple wall, and Claude's consciousness itself merge into the act of chiselling: "Claude frappait presque sans conscience [...]. Sa pensée en miettes, effondrée comme le temple, ne tressaillait plus que de l'exaltation de compter les coups: un de plus, toujours un de plus... Désagrégation de la forêt, du temple, de tout..." (*La Voie royale*, p. 82). Claude fails, however, and the jungle triumphs: "après tant d'efforts, la forêt reprenait sa puissance de prison" (p. 83), the narrator notes.

Although Claude's act lacks the political dimensions of those of other Malrucian heroes, it stems from an impulse shared by the heroes of the Asiatic novels: to reclaim a form of dignity in the confrontation with the natural world in what Vin Dao considers "la résistance de l'homme contre la déchéance et l'humiliation. [...] Être homme,

[8] Albert Camus, *Le Mythe de Sisyphe: essai sur l'absurde* (Paris: Gallimard, 1987), p. 46.

c'est résister au destin, imposer la volonté humaine face à l'univers incohérent, c'est agir au lieu de subir".[9] This confrontation therefore becomes an act of revolt echoed in Perken's desire in the same novel to "laisser une cicatrice sur cette carte" (p. 60); that is, to leave the mark of his will, a form of individual permanence, on a landscape that would deny this.

It is when this desire is located within the Chinese revolutionary context, however, that it begins to take on the dimensions of a revolt aimed more broadly at reclaiming human dignity in the face of a hostile natural world. The revolutionary action of *Les Conquérants* therefore constitutes a revolt against the apparent stasis of the natural world and an attempt to break free of the inertia that constantly threatens Malraux's revolutionary characters. Thus revolutionary violence is pitched once again against the torpor of the Asian night, but the tone here is one, initially, of hope suggested by the sound of gun fire: "La vie est collée au sol: [...] lampes à la flamme droite dans la nuit chaude et sans air, ombres rapides, silhouettes immobiles, phonographes, phonographes... Au loin, pourtant, des coups de fusil" (p. 253). Moreover, in *La Condition humaine*, revolutionary action is not so much a revolt against socio-political stasis as a revolt against the stillness of the world itself suggested by the revolutionaries who constitute "une immense foule [qui] animait cette nuit de jugement dernier" (p. 25). For Hemmerlich, therefore, the dignity of Communism lies in the hope it places in "*la force humaine en lutte contre la Terre...*" (p. 330, author's emphasis).

Revolutionary action in the Asiatic novels throws Malraux's characters into direct confrontation with the natural world. This very action also shapes their perception of the world, transforming it from an indifferent force to one that appears hostile towards human activity, denying the latter any value. In the Asiatic novels, the natural world is a reflection of cosmic time: constant, self-sufficient, and into which the Malrucian hero cannot be integrated. Revolutionary action constitutes a challenge to the inertia the Malrucian hero perceives in this permanence. It is informed by an understanding of time that elevates the value of action and therefore of engagement with the present in an attempt to forge the future. The tension experienced by Malraux's characters therefore results from a conceptualization of time in which

[9] Vin Dao, *André Malraux ou la quête de la fraternité* (Geneva: Droz, 1991), p. 41.

the present is directed by a future-orientated consciousness. This shapes the world around it into a series of obstacles, littering the ground on the path to an action that, once completed, will simply fall away to reveal the next challenge to the ever-questing Malrucian hero. The tension at the heart of the Asiatic novels is therefore born of the opposition between two different conceptualizations of time and space. It is what Bakhtin would consider a chronotopic tension; one that draws on the dialogic opposition of two opposing conceptualizations of time and space.[10]

The Anti-Fascist Works

The parallel battles between revolutionary action and the forces of political stasis on the one hand and the struggle for human dignity and the forces of nature on the other continue in Malraux's anti-fascist novels, *Le Temps du mépris* and *L'Espoir*. In both, nature remains indifferent to the presence of humanity but also, when the latter engages in action, seemingly hostile to this presence. Thus, when one enemy assault is repulsed, Manuel notes that it is:

> [...] comme si [la vague ennemie] n'eût pas été défaite par les anciens miliciens mais par la pluie éternelle qui déjà mêlait beaucoup de leurs mots [*sic*] à la terre, et renvoyait vers d'invisibles tranchées les vagues d'assaut ennemies, effilochées et dissoutes, à travers le voile de pluie aux détonations aussi nombreuses que ses gouttes.[11]

Again, the emphasis remains on nature's permanence but also on its ability to eradicate humanity's presence, to dissolve the latter into its murky indifference.

The natural world continues to be associated with the stillness of cosmic time in contrast to the Malrucian hero's own sense of finitude. Nowhere is this more evident than in the episodes involving flight. Flight in these works not only confronts Malraux's heroes with

---

[10] The chronotope is, according to Bakhtin, "the intrinsic connectedness of temporal and spatial relations that are artistically expressed in literature". M.M. Bakhtin, "Forms of Time and Chronotope in the Novel", in *The Dialogic Imagination*, ed. by Michael Holquist, trans. by Caryl Emerson and Michael Holquist (Austin: University of Texas Press), pp. 84-259 (p. 84).

[11] André Malraux, *L'Espoir* (Paris: Gallimard, 1992), p. 426.

the forces of nature, but also provides a glimpse of the cosmos in the escape from the earth itself. Thus, in *Le Temps du mépris*, it appears to the communist hero Kassner, escaping Nazi Germany in a light aircraft:

> [...] qu'ils venaient d'échapper à la gravitation, qu'ils étaient suspendus avec leur fraternité quelque part dans les mondes, accrochés au nuage dans un combat primitif, tandis que la terre et ses cachots continuaient sous eux leur course qu'ils ne croiseraient plus jamais.[12]

Similarly, in *L'Espoir*, after the successful bombing of the Nationalist barracks in Talavera, and as the Republican bomber rises above the cloud in the night sky, Malraux's aviators sense that the glimpse of the cosmic afforded by flight negates their human achievements:

> [...] aucun geste humain n'était plus à la mesure des choses; [...] l'euphorie qui suit tout combat se perdait dans une sérénité géologique, dans l'accord de la lune et de ce métal pâle [de l'avion] qui luisait comme les pierres brillent pour des millénaires sur les astres morts. (pp. 256-57)

Yet, flight, like the action of earlier Malrucian heroes, constitutes a revolt against the natural world, but in the name of human, rather than individual dignity. In the storm that engulfs his plane, Kassner and his pilot are united fraternally in a battle with the elements: "le besoin de la revanche étaient avec eux dans la carlingue contre l'ouragan" (*Le Temps du mépris*, p. 136). Consequently, the fraternity of Malraux's aviators becomes a force for countering the forces of nature. Thus, in *L'Espoir*, during the Battle of Guadalajara, the fraternal struggle of the aviators with the elements constitutes a rival force to this latter:

> L'indifférente mer de nuages n'était pas plus forte que ces avions parties aile contre aile, en vol aile contre aile vers un même ennemi, dans l'amitié comme dans la menace cachée partout sous ce ciel tranquille; que ces hommes qui acceptaient tous de mourir pour autre chose qu'eux-mêmes, unis par le mouvement des compas dans la même fatalité fraternelle. (pp. 533-34)

Flight therefore constitutes a challenge on the part of humanity to the cosmic order of things from which humanity is excluded. Thus, despite the strength of the storm into which Kassner's plane

[12] André Malraux, *Le Temps du mépris* (Paris: Gallimard, 1935), p. 134.

flies: "la constance du moteur laissait croire encore à la domination de l'homme" (p. 137). The heroes of Malraux's anti-fascist novels therefore continue to see humanity's hereditary enemy in the world around them, rather than only in their political enemies of the 1930s, as Vin Dao contends.[13]

Furthermore, the fraternal struggle against the natural world, through flight and all its dangers, endows Malraux's characters with a sense of human permanence often lacking in the Asiatic novels: in their collective struggle, and more particularly in their return to earth, Malraux's characters realise that humanity's struggle with the natural world is an age-old phenomenon, one that confers a certain permanence on human activity. Thus the view of the earth from above supplies Kassner with a glimpse of humanity's constant struggle with the earth, a struggle whose permanence not only reinstates human dignity, but which also confers permanence on humanity's presence in the world:

> De seconde en seconde entre les nuages les plus bas apparaissait et disparaissait tout l'opiniâtre monde des hommes; le combat contre la terre inépuisablement nourrie des morts et qui de minute en minute se plombait davantage, parlait à Kassner d'un accent aussi sourdement souverain que celui du cyclone rejeté en arrière; et la volonté des siens [les communistes] acharnés là-bas, [...] montait vers les derniers reflets roux du ciel avec la même voix sacrée que l'immensité – que le rythme même de la vie et de la mort. (*Le Temps du mépris*, pp. 143-44)

It is in *L'Espoir*, however, that Malraux refines this discovery of human permanence in the return to earth experienced in the celebrated descent from the mountain. Here Spanish peasants come to the rescue of one of the international aircrews shot down in the mountains above Teruel. The ascent of the mountain to retrieve the dead and the wounded is largely seen through the eyes of Magnin, the squadron leader, who initially senses the eternal rooted inaccessibly in the mountain landscape. While human activity seems to suggest a succession of events, reflected in the sight of Segunte and its fortresses where Christian, Roman and Punic ramparts have all been constructed on top of one another (p. 544), the sight of an apple tree which thrives

[13] Thus Vin Dao contends that, for Malraux's revolutionaries: "L'ennemi de l'homme ne s'appelle plus le destin, il est la société qui le condamne à vivre comme une bête", *André Malraux ou la quête de la fraternité*, p. 142.

amidst “l’indifférence géologique” (p. 549), feeding off the ring of apples rotting at its roots, suggests both the world’s indifference to human activities and its natural permanence in contrast to humanity’s fragility.

Magnin’s encounter with the local peasants reveals them to be “sans époque” (p. 560), however, possessing a timelessness that results from the struggle with this inhospitable environment. The description of the wounded then suggests a parallel between the seemingly eternal suffering of the Spanish people, suggested earlier in the novel through a series of biblical allusions likening their misery to that endured during the Exodus, and that now endured by Magnin’s aviators.[14] The wounded thus remind Magnin of “des gravures de vieux supplices […]” (p. 500), while Gardet’s disfigured face suggests “une Présentation du combat” (p. 554). This recognition of the eternity of suffering is echoed in the peasants’ recognition in Gardet of “l’image même que, depuis des siècles, les paysans se faisaient de la guerre” (p. 561).

As the cortege descends the mountain, it takes on a rhythm which itself suggests an affinity between the suffering of humanity and the eternal, cyclical movement of nature: “Et ce rythme accordé à la douleur sur un si long chemin semblait emplir la gorge immense où criaient là-haut les derniers oiseaux, comme l’eût emplie le battement solennel des tambours d’une marche funèbre” (p. 559). In the descent from the mountain, Malraux’s aviators appear to Magnin to share in nature’s eternal rhythm not through a transcendental relationship with the natural world itself, but through the recognition of what binds them to the people: the shared suffering reflected in a people engaged in an age-old struggle with the earth and thereby bound up in its eternal rhythm.

Similarly, the descent from the mountain in *Sierra de Teruel*, which virtually mirrors that of Malraux’s Spanish Civil War novel, seems to replace the characters within a continuum of eternal human suffering. This is suggested here by the visual parallels between the descent from the mountain and the descent of Jesus from the cross in

[14] See Martin Hurcombe, “The Ideology of the Early Twentieth-Century Novel as a False Ending” in *Dialogues 2: Endings* (Exeter: Elm Bank, 1999), ed. by Ann Amherst and Katherine Astbury, pp. 67-75.

Italian Renaissance art, as Marcel Oms argues.[15] As in the novel, the film evokes the peasants' intimate relationship with the world around them. This fusion of the people with their inhospitable environment is suggested in one of the final shots of the film. Here the villagers seem to barely emerge from the rock of their buildings, which themselves seem carved out of the face of the mountainside, suggesting both the villagers' struggle with the earth, but also their own rock-like permanence. The film closes on a shot of the people forming a Greek Z in the mountainside, suggesting, Oms argues, immortality.[16] Again, we sense the people emerging from the mountainside, but sharing with it something of its permanence, a permanence that is suggested in the inexorable movement of the cortege which, as Malraux wrote in his own directions, "s'étend à perte de vue".[17]

In both *L'Espoir* and *Sierra de Teruel*, the descent from the mountain following the return to earth suggests a form of human permanence through the notion of rhythm, a rhythm that is that of the funeral cortege, but which suggests the circle of life and death. The apple tree with its ring of rotting apples, which Magnin notices on his ascent of the mountain in *L'Espoir*, therefore becomes during the descent a metaphor for human existence; the death of one of Magnin's aviators is therefore no longer an individual tragedy, but integrated into a wider notion of human community, a community that travels through time, but which draws its strength to continue through its struggle with an inhospitable world it now rivals. As Raymond states: "In the descent from the mountainside, the amoral force of nature is countered by the vigour of the spirit which animates the peasants and fliers, and which establishes the parameters of the real space to be occupied by man [*sic*] in the world [...]".[18]

---

[15] For Marcel Oms, therefore, the film's greatest achievement is its "fusion de deux esthétiques, la tradition chrétienne et l'art des visages et des foules mis au point en URSS par Eisenstein et Dovjenko", *La guerre d'Espagne au cinéma* (Paris: Éditions du Cerf, 1986), p. 130.

[16] Oms, *La guerre d'Espagne au cinéma*, p. 131.

[17] André Malraux, *Espoir: Sierra de Teruel: scénario du film* (Paris: Folio, 1997), p. 165.

[18] This space, Raymond goes on to state, is "not a space captured in an all-encompassing ideological or proprietorial net, but a sense of place rooted in man [*sic*]; his compassion, moral autonomy and fraternity". *André Malraux: Politics and the Temptation of Myth*, p. 129.

In this way, the return to earth enables Malraux's heroes to discover the human community. More importantly perhaps, it allows them to gain a sense of their place in the world, not as individuals cast forever against the world as horizon, to return to Bakhtin's terminology, but as part of an encompassing environment where the present is experienced in relation to a past rather than purely in terms of a project to be completed. The Spanish people in Malraux's Spanish Civil War works therefore possess a temporal depth that results from their engagement with the natural world, but this temporal depth and the awareness of the resulting human permanence derive from the aviator's experience and perception. It is he who has encountered the force of the cosmos, Kassner in *Le Temps du mépris* and Magnin in *L'Espoir*, who is then able to bestow a sense of permanence on the human community in his subsequent encounters with the earth. In this, and as Raymond argues, Malraux's aviators are promethean, engaged in an act of revolt on behalf of humanity.[19]

The Experience of the Second World War

The anti-fascist novels adumbrate a new conception of time and space in Malraux's work. Here, the political activities of Malraux's characters begin to be understood within the framework of a constant human struggle that traverses time and is not simply the result of an immediate set of politically determined circumstances. As such, they are understood, through parallels to other cultural representations of human suffering, as aesthetic as well as political events. They are equally possessed by a rhythm that is born of the fraternal struggle for human permanence. For Bakhtin, rhythm is primarily perceived in the other, "a form of relating to the other [rather than] a form of relating to myself [...]" (p. 120). Yet, the individual experiences rhythm by participating in the rhythm of "a communal mode of existence [...]" (pp. 120-21). Rhythm therefore integrates the future-orientated activities of the individual into a broader context, translating these from the plane of the horizon to that of the encompassing environment. In the resulting reconfiguration of time and space, a solution is offered to those of Malraux's heroes previously prone to experiencing what Paul Ricœur

[19] Raymond, *André Malraux: Politics and the Temptation of Myth*, p. 182.

terms "a deficit in being" characterized by the pursuit of the eternal and the concomitant realization that it exists beyond our human experience of time.[20]

This sense of rhythm informs the depiction of the French peasantry in Malraux's final novel, *Les Noyers de l'Altenburg*. War, Vincent Berger observes as his tank regiment tear through the French countryside in May 1940, has destroyed "[le] vieil accord de l'homme et de la terre [...]".[21] The world around Berger and his comrades in arms is transformed into a horizon by the confrontation with death as they advance on German positions: "L'univers devient une indifférente menace" (p. 267). Berger and his men survive battle but only after a near-death experience in which their tank falls into a German tank trap.

Taking shelter in a nearby village, Berger discovers a world that has survived the trials and tribulations of history. The barns in which he and his comrades sleep are "les granges des temps gothiques" (p. 288). While war has apparently destroyed the accord between the peasantry and the natural world, the evidence of peasant life reveals: "la vieille race des hommes que nous avons chassée et qui n'a laissé ici que ses instruments, son linge et ses initiales sur des serviettes, [mais qui] me semble venue, à travers les millénaires, des ténèbres rencontrées cette nuit [...]" (p. 288). A peasant woman who has refused to leave the village appears to Berger: "Accotée au cosmos comme une pierre..." (p. 289).

In this notion of the accord between the peasants and the world, Malraux is not suggesting a subservience of the human to the natural, nor is he suggesting the integration of the former into the latter. Rather, he is suggesting a rhythm that is predicated upon that of the natural world and that is born of the conflict with the latter. This is suggested in the walnut trees of the novel's title. These trees, Berger's father realizes, possess a rootedness that allows them to traverse time, drawing their force from the earth to which they so desperately cling:

> La plénitude des arbres séculaires émanait de leur masse, mais l'effort par quoi sortaient de leurs énormes troncs les branches tordues, l'épanouissement de feuilles sombres de ce bois, si vieux et si lourd qu'il semblait s'enfoncer dans la terre et non

---

[20] Paul Ricœur, *Time and Narrative*, vol. 1, trans. by Kathleen McLaughlin and David Pellauer, (Chicago: University of Chicago Press, 1984), p. 26.
[21] André Malraux, *Les Noyers de l'Altenburg* (Paris: Gallimard, 1948), p. 269.

> s'en arracher, imposaient à la fois l'idée d'une volonté et d'une métamorphose sans fin. (p. 151)

It is his son, however, who, like Magnin in *L'Espoir*, and following his near-death experience in 1940, will realise that the tree is suggestive of humanity's collective and tenacious struggle for permanence in an inhospitable world.

Large parts of *Les Noyers de l'Altenburg* are inspired by Malraux's own wartime experiences and are therefore reproduced in his anti-autobiography *Antimémoires*. Like Vincent Berger, Malraux served in a tank regiment in 1940 and was captured by the Germans. He would go on to fight in the Resistance, again being captured by the Germans in 1944, and then, after the Liberation, lead a regiment of the French army. It was perhaps by choosing tanks over aviation in 1940, before joining the *maquis*, a form of Resistance whose very name suggests a confusion with the soil of France, that Malraux, the erstwhile aviator of the España squadron in Republican Spain, came back down to earth himself and encountered the seemingly timeless French people.[22]

In *Antimémoires*, this encounter occurs primarily following Malraux's capture in 1944. Imprisonment at the hands of the German forces places Malraux in the position faced by a number of his characters (Kyo and Kassner, to give but two examples). Here he feels: "amputé de l'éternel" due to the imminence of his own death.[23] Yet the support afforded him by the local population as he approaches his death supplies a form of fraternity that in turn supplies a sense of existing not as an individual subject delimited by personal finitude, but as part of an encompassing human environment that will survive this:

> Mon passé, ma vie biographique n'avaient aucune importance. […] Je pensais aux paysannes athées qui saluaient mes blessures du signe de la croix, à la canne apportée par le paysan craintif, au café de l'hôtel de France et à celui de la Supérieure. Il ne restait dans ma mémoire que la fraternité. Dans ce silence de couvent où sans doute on priait pour moi […], ce qui vivait aussi profondément en

---

[22] Janine Mossuz-Lavau therefore asserts vis-à-vis Malraux's Resistance experiences: "[La France] a gagné dans cette épreuve une réalité charnelle qui n'apparaissait guère quelques années plus tôt". Janine Mossuz-Lavau, *André Malraux et le gaullisme* (Paris: Presses de la fondation nationale des sciences politiques, 1982), p. 31.

[23] André Malraux, *Antimémoires* (Paris: Gallimard, 1967), p. 229.

> moi que l'approche de la mort, c'était la caresse désespérée qui ferme les yeux des morts.[24]

Malraux's comfort in the face of death therefore lies in his ability to lose himself however briefly in the rhythm of human community.

In *Antimémoires* the reproduction of Second World War episodes from *Les Noyers de l'Altenburg* alongside memoirs of Malraux's ministerial travels to India, the French West Indies, and China reveals an interest in the national community as a human continuum into which the death of individual members is rhythmically integrated through national memory and ritual. Thus the funeral pyres that burn on the banks of the Ganges, surrounded by a ring of mourners, remind Malraux of "les files qui montèrent lentement vers les bombardements, par la Voie sacrée de Verdun, par la route de Stalingrad" (p. 265), but also, in a reference to *Les Noyers de l'Altenburg*, of "l'anneau de noix mortes, là-bas, en Alsace, autour d'un tronc noueux – comme cet anneau de vivants autour d'un corps qui semblent brûler à regret" (p. 272). Like the tree feeding off its own fruit, the nation draws its strength from its own dead. Once again, nature reflects human permanence. The nation, like the fraternal antifascist community of Malraux's works of the late 1930s, therefore comes to constitute a form of community that traverses time.

As Geoffrey Harris observes with regard to *Antimémoires*, there is, in the dialogue of past and present, of fact and fiction, "a neo-Proustian resurrection of the past through art as Malraux occults spatial and temporal barriers".[25] *Antimémoires* reflects Malraux's continued pursuit of human persistence through the study of the human community in its many national forms, drawing on a selection of writings that span his literary, military, and political career, integrating these many national forms into a global human community, often facilitated by flight as Malraux the aviator becomes Malraux the minister travelling the world in the name of the Republic. The French nation thus becomes in Malraux's thinking "un personnage surnaturel", the Republic: "l'intercesseur entre la vie humaine et le monde inconnu, entre la misère présente et le bonheur futur, et d'abord entre la solitude et la fraternité" (p. 169). In such a nation, de Gaulle fulfils the promethean role of Malraux's aviators, interceding

---

[24] André Malraux, *Antimémoires*, pp. 230-31.
[25] Harris, *André Malraux: A Reassessment*, p. 213.

between the nation and its destiny, since he is "un personnage hanté, dont ce destin *qu'il devait découvrir et affirmer* emplissait l'esprit. Chez un religieux: la personne, le sacerdoce, la transcendance" (p. 135). Malraux's *Antimémoires* therefore suggests that his adoption of the politics of Gaullist nationalism derives from a movement that is already discernible, ironically, in those works most closely associated with the politics of the French left. The substitution of the national community for the fraternal revolutionary community is predicated upon a need shared by Malraux's characters and by Malraux the essayist and politician for a sense of human permanence in a world that would deny this.

The nation therefore supplies Malraux with another example of a human environment that, through its struggle for meaning in a world inherently devoid of this, gains a sense of its own permanence and becomes attuned to cosmic time. The latter, as Malraux writes in *Antimémoires*, whilst bound in a constant cycle, is not immune to change. In this way, he distinguishes eternity from cosmic time. The latter is: "ce temps *animé* par la naissance, la vie et la mort de ses cycles, [qui] entre dans une dialectique sans fin avec l'essence du monde, qui ne renaîtra point semblable à ce qu'elle est – malgré l'inéluctable retour à son origine éternelle" (p. 267). The nation derives its strength from its past, but will change with time whilst remaining true to itself; a key tenet of Gaullist thought that allowed for the modernization of France and the defence of national hegemony throughout some of the coldest years of the Cold War.

For Malraux, the nation is not an organic, natural entity, as it was for Maurice Barrès, for example; there is no human integration into the natural order. As much as they may echo nature's permanence, Malraux's communities, like the Spanish peasants at the end of *Sierra de Teruel*, stand apart from the natural world whilst remaining attuned to it. The reconfiguration of time and space through the discovery of permanence in humanity's eternal struggle with the natural world, and which follows the return to earth, allows for the creation of what might be termed a second nature: a world mirroring cosmic time, but essentially a human creation in a meaningless universe. The contingency of human existence in relation to the natural world therefore remains a constant throughout the works discussed here. While the chronotopic tension observed in the Asiatic novels is eased by Malraux's reconfiguration of time and space, it is never entirely eradicated. Yet, in the course of the 1930s Malraux's conception of the world undergoes a radical transformation; through the discovery of fraternity and of human community, the world is transformed from the antagonistic horizon of the Malrucian adventurer to the encompassing environment of *l'homme fondamental*. The national community is therefore the final incarnation in a series of communities whose origins are to be found in those works often associated with revolutionary activity, but whose very existence responds to a dilemma underlying all of Malraux's artistic output.

Index of Names